Safety Rebels

Safety Rebels: Real-World Transformations in Health and Safety discusses the pragmatic experiences of over 30 safety professionals worldwide who managed to positively transform safety within their organizations. This book details the approaches taken while considering the politics and dynamics within each organization, including resistance to change, deteriorating safety statistics, increased number of procedures for operational personnel, high turnover, and budget restrictions.

Professionals from the world of aerospace, healthcare, energy, oil, rail, and public services share their experiences of positive safety change, revealing transformations in different contexts. This book explains key safety principles, theories, and shared models. It reveals how the professionals applied state-of-the-art knowledge, Safety-II, FRAM, incident data, and artificial intelligence into their organization to prevent personnel from working in a silo. It includes relevant safety and multidisciplinary theories, from Heinrich's incident model to resilience engineering, as well as aspects of change management and human organizational performance (HOP). These theories offer the reader a framework to try something new, and this book will inspire them to try contemporary strategies and tactics to approach safety challenges within any organization.

This timely and easy-to-read book will appeal to professionals in the field of health and safety. It will have particular appeal to those working in Industrial Engineering, Aerospace Engineering, Nuclear Engineering, Mechanical Engineering, Civil Engineering, Chemical Engineering, Biomedical Engineering, and Electrical Engineering.

Selma Pirić is an independent safety expert and trainer for her company, Human Performance Training & Consultancy. She started her career in airline operations, and since then, she has worked in various international aviation consulting roles in the UK and the Netherlands, as a certified university lecturer and researcher, and has worked at the Dutch air traffic control as a safety expert.

Workplace Insights: Real-World Health, Safety, Wellbeing and Human Performance Cases

Series Editor:
Nektarios Karanikas and Sara Pazell

The aim of the series is to host and disseminate real-world case studies at workplaces with a focus on balancing technical information. Further, the application of a work design framework will propel this series into the literary cross-over of traditional occupational health, safety, or wellbeing, human factors engineering, or organizational sciences, into a design realm like no other series has done. Each case will describe the tools and approaches applied per the Work (Re)Design stages and inform the readers with a complete picture and comprehensive understanding of the what's and why's of successful and "failed" attempts to improve the work health, safety, wellbeing, and performance within organizations.

Safety Insights
Success and Failure Stories of Practitioners
Edited by Nektarios Karanikas and Maria Mikela Chatzimichailidou

Ergonomic Insights
Successes and Failures of Work Design
Edited by Nektarios Karanikas and Sara Pazell

Healthcare Insights
The Voice of the Consumer, the Provider, and the Work Design Strategist
Edited by Sara Pazell and Jo Boylan

Safety Rebels
Real-World Transformations in Health and Safety
Selma Pirić

For more information on this series, please visit: https://www.routledge.com/Workplace-Insights/book-series/CRCWIRWHSWHPC

Safety Rebels
Real-World Transformations in Health and Safety

Selma Pirić

CRC Press
Taylor & Francis Group
Boca Raton London New York

CRC Press is an imprint of the
Taylor & Francis Group, an **informa** business

Designed cover image: © Shutterstock

First edition published 2024
by CRC Press
2385 NW Executive Center Drive, Suite 320, Boca Raton FL 33431

and by CRC Press
4 Park Square, Milton Park, Abingdon, Oxon, OX14 4RN

CRC Press is an imprint of Taylor & Francis Group, LLC

ISBN: 9781032467542 (hbk)
ISBN: 9781032464459 (pbk)
ISBN: 9781003383109 (ebk)

DOI: 10.1201/9781003383109

Typeset in Times
by codeMantra

Contents

Foreword xi
Preface.. xiii
Acknowledgments.. xv

Chapter 1 The Only Constant in Safety Is Change.. 1

The Courage to Change – A Tale of Taking Risks 1
    Ask for Forgiveness, Not for Permission... 3
The Oil Slick Approach to Promoting Safety 3
    Lessons from the High Seas 5
    The Ripple Effect: Small Changes, Big Impact 5
From Compliance to Compassion: The Organizational Learning
Potential.. 6
    Focusing on High Learning Potential to Overcome
        Safety Issues 6
    Creating Learning Potential through Leadership............................ 7
    Language and Shaping Stories from Incidents................................ 9
    'Don't Lose Situational Awareness!' 9
    Reimagining Safety: Through the Lens of Trust........................... 10
The Only Constant in Life is Change... 11
Conclusion 11
References 12

Chapter 2 The Good, the Bad, and the Ugly: The Evolution of Safety 14

Introduction 14
The Ugly: The Delta Incident... 14
    Incident Development.. 15
    Incident Analysis 16
The Bad: The Criminal Prosecution 16
The Ugly – Again – The Aftermath.. 17
The Good: Lessons Learned 19
Learning That Transcends Organizational and National Culture...... 19
Protecting the Reporters... 20
    Reporting: Why One Would or Wouldn't....................................... 21
    The Need for Speaking Up Versus the Need to
        Not Look Stupid.. 21
    Ask the Trainees Rather than Telling Them What to Do.............. 22
    A Parallel Development: Psychological Safety 25
Conclusion 26
References 26

Chapter 3   Safety Rebels Always Ring Twice 28

Introduction 28
Organizing Work in the Anglo-Saxon or Rhineland Approach 28
    Safety Training ≠ Guaranteed Safety Improvement 30
    Similarities in Managing Modern Society
        and Managing Safety 30
    A Collective Ambition toward Organizational Goals 31
    The Confined Chicken 32
    Complex Versus Complicated 33
Managing Safety 33
The S-Word 34
    Problems with Safety Culture 35
    Climbing the 'Ladder' 35
    The First Rule of Safety Culture 36
    Safety Culture: Measurement and Limitations 37
    Planning (for Safety Culture) Prevents Poor Performance 38
Conclusion 40
References 40

Chapter 4   Counting Definitely Counts, But Not Always 42

The Double-Edged Sword of Safety Performance Indicators 42
There Is No Such Thing as a Reliable Safety
Performance Indicator 42
    Outsource the Complexity of Safety Metrics to
        a Party that Understands Them 43
    Accountability Comes with Autonomy 45
    Safety Performance Indicators: A Useful Gauge
        If You Know What You're Doing 46
Gaming the System: Tipp-ex Incidents 47
    Insult over Injury 47
Unmasking the True Picture of Safety 47
    Shifting the Focus: From Safety Metrics
        to Proactive Safety 49
    Expect Resistance 50
    The Catch-22: Are We Feeding the System, or
        Is the System Helping Us? 51
    A Litigious Society: Who Do We Keep in Mind When
        Designing the System? 52
    The Subconscious and Safety Behavior 52
    Interventions: Local versus Global 53
Defining the Goals of Incident Investigation 54
    Finding Local Rationality in the Development of the Event 55
    Define the Learning from Incidents Process 56
Begin with the End in Mind 57

Conclusion 58
References 58

Chapter 5   First Rule of Leadership: Everything Is Your Fault 59

Accidentally Effective Safety Leadership 59
Setting Clear Expectations from Leaders for Safety Practices 60
But Should All Managers Have Operational Experience? 61
Prioritizing Safety Alongside Operational
and Commercial Goals 62
Aligning All Management Levels with the Safety Goals 63
Emergent Behavior from Societal Developments 63
Servant Leadership: Empowering Employees with Growth
and Just Culture 63
Designing Out Risks: Setting Challenging Operational Goals 65
Assessing Risks in a Live Operation 66
Doing the Same = Getting the Same Results 67
Aligning Personnel through Communication: Heads,
Hearts, and Hands 67
Patience, Persistence, Positivity: Keys to Effecting Change
in Organizations 68
Looking Beyond the Known: Cross-Industry Insights
and Gaining Fresh Perspectives on Safety 69
Rebranding Safety: Engaging Employees in a New Narrative 70
Conclusion 71
References 71

Chapter 6   Remember: You Have Two Ears 72

Introduction 72
Reframing the Conversation: How to Make Safety
a Priority in the Workplace 72
Opportunities for Improving Interpersonal
Communication in the Workplace 73
Creating a Profound Understanding of Work at All Levels 74
Creating Company Success Factors Everyone Believes In 74
Ditching the Safety Beating Stick 75
When Safety and Operational Goals Paradoxically Align 76
Subgoal-Criteria Were Met – But the Job Could Not Be Done 76
Explain It to Me Like I'm 5 77
A Closer Look at Leadership Behavior, Team Norms,
and Individual Personality Traits 77
Disarm, Listen, Care 79
Prioritizing a Learner Mindset 79
Lifesaving Rules Might Not Always Be Lifesaving 80
Using Unknown Hazards to Ensure Double-Loop Learning 80

Overcoming Resistance to Change in the Workplace 82
Conclusion 82
References 82

Chapter 7   Blending Theoretic Principles to Practical Advantages 83

Models in Safety and Risk Management: An Introduction 83
Patient Safety: A Growing Priority in
Health Care Organizations 84
How Healthcare Claims Have Initiated a Positive Change 85
    A Plateau in Growth: Supplementing Traditional
        Methods for Understanding What's Happening 86
    When Everything Is Urgent, Nothing Is Urgent 87
    Developing Staff 'Soft Skills' Capabilities to
        Improve Patient Safety 88
    Examining the Role of Leadership in Patient
        Safety Initiatives 89
    Future Strategies for Ensuring Patient Safety and Quality 90
Normalizing Vulnerability 91
Conclusion 92
References 92

Chapter 8   Sell Safety Like Hotcakes 93

Human-Centered Thinking as the Basis
for Safety Improvement 93
    Human Organizational Performance (HOP):
        A New Way of Learning 93
    Ticking the 'Human-Error' Box 94
    Does the Worker Fit the Job, or Does the Job Fit the Human? 95
    Selling Safety: Changing the Way We Speak about Safety 95
    The Power of Operational Learning: Insights Directly
        from the Frontlines 96
Why Is it So Difficult to Create More 'Awareness'? 97
Success in Navigating Change Is Only as Good as You Sell It 98
The Power of Telling Stories in Safety Communication 98
Simplifying Safety Documentation for SMS Improvement 100
Getting Rid of the 'Golden Rules' 100
A New Perspective on Auditing 101
Generational Change: The Long Road to Lasting Safety
Improvements 101
Conclusion 103
Reference 103

Chapter 9 Chance Favors the Prepared.. 104

Safety May Be Subjective; Risks Are Real..................................... 104
Why We Shouldn't Only Be Talking About Safety But Also
About Business Continuity 104
Change Your Safety Game: Measure the Organization's
Planning for Safety Culture... 105
Concluding Remarks 107
Reference 107

Index 109

Foreword

Alanna Ball

Safety, in both concept and practice, has always been a fundamental necessity for the development and sustenance of civilizations. It underpins our actions, structures our societies, and influences the most basic human decisions. The history of safety, how it evolves and changes is linked to our growth as a species. However, like all evolutionary processes, safety as a concept has not been static. It has transformed and expanded, oftentimes sparked by those who dared to think differently. These are the 'Safety Rebels.'

Safety Rebels is not just a chronicle of health and safety theories but a celebration of revolutionary thinkers who questioned the status quo. They recognized that safety was not merely about rules and regulations but understanding human factors, system dynamics, and an organization's culture. Exploring the 'good, bad, and ugly' to the ever-changing acronyms and systems we use to 'measure' success.

Highly reliable organizations (HROs) and industries like aviation, nuclear power, and healthcare have often been at the forefront of these evolving safety paradigms. Traditional safety paradigms focused extensively on retrospective learning from past errors, attempting to pre-empt and prevent future incidents through stricter regulations. While this approach had its merits, our safety rebels realized that in a complex and constantly changing world, merely trying to prevent past mistakes was not enough. The chapters seamlessly weave in the change, the theory, and the practicality of how we 'do' health and safety.

Reading *Safety Rebels* is like setting out on an adventurous voyage. It is a voyage that highlights the audacity of a select group of thinkers who, over time, have made significant contributions to changing the worldwide dialogue on safety. But perhaps more importantly, each chapter serves as a reminder. A reminder to everyone that evolution and progress, particularly those in the sphere of safety, frequently require rebellious thought. It needs people who are willing to challenge, question, and disrupt the status quo. It needs the courage to innovate and the fortitude to deal with skepticism.

I strongly advise the reader to view this book as an inspiration rather than merely a collection of theories. Let it take you on the journey, alter your perceptions, spark new ideas, and motivate you to think in new ways. These safety theories are more important than ever in today's interconnected, complicated world where complex systems frequently house both problems and answers. They point us in the direction of a time when safety isn't just a box to tick but a fundamental aspect of how we conduct business.

This book is an homage to your vision, to the rebels who dared to think differently, to those who overcame opposition yet persisted, and to everyone who helped shape the ever-changing terrain of safety. And to those reading *Safety Rebels*, I hope you'll be motivated to follow in the footsteps of these thinkers.

Preface

This book is for anyone interested in reading about how safety professionals around the world have taken on challenges to make positive improvements. It is particularly relevant for safety professionals and managers responsible for creating and maintaining safe working environments, as well as employees who want to better understand the importance of safety and their role in promoting it. Additionally, this book may be helpful for policymakers, regulators, and other stakeholders involved in shaping safety policies and standards. Whether you are a seasoned safety expert or just starting your safety journey, this book provides valuable insights and practical strategies to help you enhance safety performance and achieve better outcomes.

What started as an idea out of curiosity has led me to be able to speak to some of the most influential, inspiring, and incredibly knowledgeable professionals. This book has connected me with them; they have inspired, educated, and united me with them and their cause. If anything, I hope this book will achieve the same with the readers, at least to some degree. It is not about safety officers walking around telling everybody what to do and what not to do. It is an exciting field to work in because we, as safety professionals, genuinely get to improve other people's working lives to some degree. Sometimes, however, the excitement of safety becomes hidden behind other elements, such as paperwork, compliance, rules, procedures, or political games. Safety sometimes gets lost in the force field in between various stakes, stakes of operational efficiency, budgets, or other interests.

I've spoken to well over 30 industry professionals to establish the following:

- To what degree do they use existing knowledge in their day-to-day job?
- Which approaches seem to generate lasting positive results?
- How do they apply theoretical knowledge in their jobs?

Their stories have shaped this book in two ways:

- Practical approaches to enable positive safety change are not meant to be directive but to inspire the reader as it might also work in their organization.
- The theories, models, and concepts they speak about will be explained so that the reader is up to date with current hot topics in safety.

WHAT'S A SAFETY REBEL?

The Cambridge Dictionary describes the word rebel as "a person who doesn't like rules or authority and shows this by behaving differently from most people in society."

Now, I'm not any different from the next person in thoughts and behavior, and it's not that I don't like rules or authority. And neither do the professionals I've spoken with for this book. But we have one thing in common: questions of "what else can we do to improve safety? What else can we do to understand safety better? What other extra steps can we make?"

In my view, a safety rebel is an individual who challenges traditional norms and approaches to safety management in the workplace. They are not content with following established rules and regulations but instead seek to push the boundaries and find new ways to improve safety standards.

While a safety rebel's approach may sometimes be seen as unconventional or even controversial, their goal is always the same: to create a safer workplace environment for everyone involved.

I was eager to understand what safety professionals have *pragmatically* done to improve safety. Which theories, models, or principles have they applied? Have they used cross-disciplinary knowledge, perhaps from Psychology or Change Management? Did they come with something boot-strapped, and how has their approach worked out?

WHAT THIS BOOK ISN'T

This book does not pretend to be introducing *new* ideas or concepts. It merely presents what's already out there and how these ideas are used in practice.

This book is also not a scientific or social experiment. It represents what industry professionals have shared in their journey of making a positive safety change within their organization.

This book does not hold the ultimate truth. The interviews serve as examples of what may (or may not) have worked in making a change. This book serves as a tool to share these practices, not to present them as principles.

THIS BOOK IS FOR YOU IF...

- You're interested to read how industry professionals around the globe are shaping safety improvements in their fields.
- You're interested to know what current hot topics are in safety.
- You're interested in the field of safety, and you want to get up to speed on current practices in safety.
- You've just graduated, and you want to know what the real world looks like safety-wise.
- You have been working in the safety field for a while and want to have an overview of the most used theories, models, and concepts.
- You know your way in your industry, but you're interested to read what safety developments occur in other industries.
- You know nothing about safety, but you would like to know more.

I have realized that I've only just started exploring and that there is so much more to explore. I wanted to bundle these stories and share them with other professionals. And that's how this book was born. I firmly believe that change is inevitable, also in the field of safety. Stagnation means decline; no matter how convinced one is, the current system is perfect as it is. Yes, initially, it might be perfect, until it isn't anymore, as the only constant in life is change.

Acknowledgments

First and foremost, I would like to thank all the professionals who have shared their compelling stories with me that brought life to this book. Without them taking the time to speak to me, this book wouldn't exist. A special thanks to some of my former colleagues I now consider my friends: Nektarios Karanikas, for your guidance, advice, and wisdom and for giving me confidence in this profession; Sannie Bombeeck, for mentioning my name in rooms I'm not in; and Renée Prang, for your endless support. Furthermore, I am grateful to all the reviewers who have provided me with valuable feedback and therefore improved this book. Last but certainly not least, my husband and family for supporting me in various ways – from mental support to babysitting so I could write; and my friends for listening to my whining about the writing process. Without any of you, I wouldn't be where I am today.

1 The Only Constant in Safety Is Change

Only dead fish flow with the stream; the live fish swim against the stream.

– Danish saying

THE COURAGE TO CHANGE – A TALE OF TAKING RISKS

Have you ever been asked to present safety information to your colleagues at work? Maybe you had to explain the safety dashboard from the previous month to senior management or share a success story with other departments. If you're like me, you likely gathered all the essential information and made a PowerPoint presentation. During the presentation, there might have been a few questions, and you may have talked about what safety is, what it should be like, or what it should look like.

Well, perhaps many of us, but not Jaap van den Berg, Safety Culture Manager in aviation.

I arranged to meet Jaap at Schiphol Airport on a warm September day. We chose a bar where we could talk without interruption. When I arrived, the bar was empty, and I could hear aircraft landing, but I couldn't see them from where I was sitting. Jaap arrives dressed in business attire, exuding a professional demeanor rarely seen nowadays except among top-level management.

'What do you do when the Chief Operations Officer at a major airline in Europe gives you the task of getting *safety between the ears of 30,000 employees*?' he asks.

Sounding like an impossible task, I struggle to place myself in his shoes and take on this enormous responsibility.

'You say *consider it done!*' Jaap says.

I admire this can-do attitude. It takes much effort to manage the safety of such a large workforce with diverse professions and skills. Safety management in an airline is a complex task beyond just managing flight operations. There are also the maintenance department, the cargo department, the baggage department, the people on the platform, the people behind the check-in desks, office staff, the many other essential airline workers I need to mention, etc.

How do you get the *message of safety* across within an organization with so many facets of operating the business?

'Safety first' is a mantra we are all well familiar with. Jaap explains that within the airline, similar mantras were repeated regularly, such as '*Help and challenge each other when you see something unsafe*' and '*Report unsafe situations.*' These mantras were put in writing, printed on posters, and communicated throughout the organization in various ways. But how do you determine whether people still hear and live these messages? How would you even measure this?

DOI: 10.1201/9781003383109-1

One morning, when Jaap received a call from the head of Human Resources (HR), they asked him if he could come to a top management meeting in the following week because the executive meeting needed *'also to include something about safety.'* Mind the general vagueness and implicit request, it would probably have to be something positive the airline has achieved recently.

He only needed to know that the program was already more or less complete, so his contribution should take little time. Safety professionals see this all the time. 'Say something nice about our safety, and then please leave.'

'Okay,' he said,

'I can do it, but it will not be a PowerPoint presentation.'

'What is it then?' HR asked.

Jaap talked them through his risky plan.

'Fine,' said HR.

'Get in touch with the event coordinator, and they will help you organize it.'

This sounds like a firm agreement, although HR probably hoped the event coordinator could not arrange it.

When Jaap contacted them, the event coordinator said 'no' immediately; it would not be possible. There would be many high-profile top managers, including directors, and they could not imagine Jaap's plan working out.

'Tell you what,' Jaap proposed.

'I will go to the location of the meeting right now and ask the organizers whether it can be arranged, and if they say it is possible, we will do it. If they say it cannot, I will be honest with you and skip it.'

The event coordinator agreed to the proposal, possibly thinking the location organizer wouldn't approve it. Jaap visited the location, and the location organizer found the plan reasonable, with no obstacles preventing its execution. Jaap informed the event coordinator that the presentation would proceed and that he would carry out the plan. Soon after, however, they began to withdraw.

'No, no! We had second thoughts about it, and we do not want to do it; it cannot take place!' the event coordinator replied.

'No,' said Jaap.

'This is what we agreed, I am going to do it, and only I am the one to blame.'

Before the meeting started, Jaap went to the location, creating several 'unsafe situations.' First, at the coffee table with bottles of juice and water, he planted three boxes of 'rat poison' (the boxes he made himself), complete with biohazard logos, skulls, and warning signs. Then, he planted various unattended bags throughout the meeting room. Moreover, while the people were starting to walk into the room, he was standing on a 4-meter ladder, in overalls, looking rather ragged and unshaven, with a cap backward on his head and sunglasses, pulling some cables from the ceiling. It could have been a scene from the movie 'Ocean's Eleven.'

Jaap hoped the management team would notice the potential hazards and raise concerns. They could have informed the staff or inquired about his activity on the ladder. Unfortunately, no one questioned the situation. Despite the presence of top and safety management, no one checked if his actions were safe or inquired about the rat poison boxes placed near the coffee machine. Nobody addressed the unattended bags or the overall safety of the environment.

Then, the chair of the event started the program, mentioning the usual housekeeping, such as where the emergency exits were, safety first, etc.

'By the way, someone here would like to say something about safety.'

Jaap descended the ladder and took to the stage, still wearing his overall. He began discussing the company's values regarding addressing unsafe situations, reporting them, and prioritizing safety. The audience appeared to agree with the importance of these values. As Jaap continued speaking, he removed his backward cap and undid his overalls, revealing himself as a corporate suit-clad business professional, ready for serious business. This sudden transformation caused the room to realize that they had not reported any unsafe situations, despite this being the expected behavior from frontline employees. Jaap's actions spoke louder than any presentation slideshow could, making the top-level management painfully aware of their responsibilities. Despite the initial disruption, Jaap received feedback on his impact. Although everyone in the room had witnessed the situation, no one had reported it.

> The room full of top-level management was made painfully aware of what *they* were constantly repeating, but did not do themselves.
>
> – *Jaap van den Berg*

ASK FOR FORGIVENESS, NOT FOR PERMISSION

If Jaap had briefed Human Resources on what he planned for this meeting, he would have never gotten permission to do it. When you are only given a brief moment to 'say something about safety,' it is unfortunate that safety is usually seen as another tick of the box. Nevertheless, through his PowerPoint-less 'presentation,' he made a powerful impact, which Jaap's manager acknowledged afterward.

Some things do not follow organizational processes or compliance tick boxes. For example, if you are tasked to get a safety message across to 30,000 employees as diverse as society itself, only publishing a safety manual is not going to cut the chase; neither is a safety portal nor using a thousand abbreviations no one understands. Communicating about safety always seems to be a balancing act; if it is focused too much on compliance, it will not get through to the employees. At the same time, if you use humor, people might not take safety seriously or get upset that you mock safety. It is not about ridiculing the safety policy, but you can show how challenging your role as a safety professional can be and use that vulnerability to engage with people. Managing change is a long process filled with hurdles and takes significant effort from all stakeholders.

Small changes might be the catalyst for acceptance among even the most skeptical opponents of change. Sometimes, all it takes to start a change is to get one group of employees enthusiastic in the organization.

THE OIL SLICK APPROACH TO PROMOTING SAFETY

In maritime, every ship has its own culture aboard, even if they all fall under the same organization. I spoke to a Quality, Health, Safety & Environment (QHSE) manager in the oil and gas industry, and he explained that ships can be the home to hundreds of workers. Operating multiple vessels means hosting multiple different cultures. It is like living in different villages.

The QHSE manager's experience includes working for a marine contractor that operates various vessels for various purposes. The QHSE manager was responsible for safety, health, and the environment in one of these maritime organizations (let's give this organization a fictitious name, *The Marine Contractors*). The Marine Contractors organization took on projects from clients such as oil companies and offshore businesses.

A client assigned one of their ships (let's call it *Ship 1*) to a new project. The clients' requirement for this project was to follow their program of behavioral change management. This was familiar because when there was a new client and a new project, it usually meant new procedures or new rules. This program was not about rules, regulations, and procedures. Instead, the program entailed addressing each other about unsafe situations and watching out for each other.

Typically, this kind of training is not received with enthusiasm, as it addresses the attitude toward safety and not compliance with safety. However, Ship 1's crew embarked on the program and showed great enthusiasm.

The pillars taught during the program ensured that the crew could always address each other if the situation required them to say something. The crew did not say, '*You are not following the rules!*' they started saying things like, '*I see that what you are doing might be dangerous; maybe there is another way to do it?*'. The effect transgressed further than initially intended. The program enabled crews and staff to see that safety is a collective goal, and it was about more than following the procedures and rules for the sake of it. It enabled the low-rank crew to even address the ship's captain if they saw them do something unsafe. Although the process took years, the barrier to addressing each other was lowered and accepted by all.

When management stepped on board Ship 1 to see the program's effect on the ship's culture, they saw a remarkable positive change in safety behavior. Machinery was labeled with warning stickers, the crew walked around in appropriate safety equipment, and the crew addressed each other. Safety-wise, it seemed as if this ship was always one step ahead of the rest of the ships.

The differences between Ship 1 and the other ships suddenly became quite visible. The organization faced a challenge: How should the new attitudes be transferred to the other ships? It was clear that forcing people to follow a particular program was not an option because it probably would not lead to the wanted change in behavior.

Consequently, the other crews started to ridicule Ship 1's crew, calling them names and asking whether they needed this to be able to do their demanding jobs, stating that these types of programs do not fit within their line of work. Furthermore, other crews thought it did not fit with who they were as hard-as-nails offshore professionals. Still, management decided to roll this program out on all the organization's ships. The program was adapted fleetwide.

Sending offshore workers to safety training is challenging, as they are tough guys and girls. They question whether they need it in their line of work.

– A QHSE manager in the oil and gas industry

Lessons from the High Seas

The responsible departments looked at the various levels of seniority of crews on Ship 1. Each crew consists of various levels, ranging from low to high rank, and crew members can be changed between ships if they fill the same rank on the other ship. The organization decided to have members of the crew that had followed the program rotate in small teams between the other ships of the organization. They picked the members of the crew they knew would have an influence and could make this impact on other ships – somewhat in the form of 'infiltration.'

The result of this effort was impressive, and it became apparent when one of the most senior crew members of the worst-performing ship in safety behaviors was holding his retirement speech. He said that one of the best things in his career was the culture change, which eventually led him to have more fun in his work. What is noteworthy is that this particular crew member was initially one of the biggest opponents of the program.

Crew members and managers worked in conjunction to make this improvement possible. In the program, there were steering teams at senior management and satellite steering teams on the ships, and these teams kept supervision over the program. In addition, the steering teams enabled taking lessons learned from the operational personnel back to the steering teams at the management level. The result was that the program was not solely a top-down initiative but also bottom-up by listening to the crew.

New hires were required to follow this program on their first day. The company made a big point of ensuring they knew how important safety was to them. When the crew came back from their five-week leave (roster rotated five weeks aboard the ship, five weeks leave), again, they received refresher training for the program. The result was not only that staff were improving safety attitudes throughout the whole organization – from offshore to office – but they also took these values back home. Staff started thinking about safety and applied it in everyday life.

The Ripple Effect: Small Changes, Big Impact

After the program was rolled out, what also became visible was that the number of reports filed by the crew was increasing sharply. Initially, it looked like safety was becoming worse, but what was happening was that the culture became more open. Incidents happened before the program started but were not reported; people did not talk about them. The organizational learning capabilities increased immensely because of this development. Additionally, the culture became less and less about blaming people for their mistakes. That was considered a miracle because when the QHSE manager started, the company had suffered a significant incident the year before, which led to punishments all around; three people got demoted and did not receive their bonuses. Other employees realized this would happen if you made a mistake.

This was now a thing of the past. Eventually, the reports that were filed gradually decreased over time. To make these positive changes last, the company repeated the training. The organization followed the complete training again every two to three years but then with different focus points, themes, and cases.

The company's focus now was on more than safety anymore; they wanted to provide the best possible quality of service to be the most reliable partner in their client's projects. The company grew to be ahead of its competition, so much so that it came in the position of being able to demand safety-related requirements from its clients. Clients were now following the same training as per the company's demand! This was, of course, absolutely the opposite way of working a few years earlier.

FROM COMPLIANCE TO COMPASSION: THE ORGANIZATIONAL LEARNING POTENTIAL

When it comes to organizational learning, organizations are inclined to investigate incidents, mainly on what went wrong and even more so if it nearly went very wrong.

'Let's not investigate based on severity, but on *learning potential*,' Barnaby Annan, Human Performance Advisor at an Integrated Energy Company, says.

Looking at learning potential means zooming in on high-consequence tasks. High-consequence tasks can significantly impact an organization's performance and safety if not done correctly. These tasks generally involve a high level of risk, require specialized skills and knowledge, or involve complex procedures with strict adherence to rules and regulations. For example, high-consequence tasks include performing critical repairs at sea, operating hazardous machinery, and any jobs involving hazardous materials.

Barnaby continues, 'The second thing you need is the resources to have meaningful conversations with the people doing that high-consequence task, including resources and toolkits.'

A new tool has been created by the Integrated Energy Company that acts as a hybrid between a learning team and a human error analysis tool. Its practical use is to determine if a gauge has been placed in the wrong location, if the alarm system is designed poorly, or if a procedure is incorrect. This tool allows for the identification and local correction of these problems.

Focusing on High Learning Potential to Overcome Safety Issues

'But what we try to do is generate more capacity to look further back in the system and ask ourselves: *How did we even end up with this in the first place?*'

Many people believe that if problems are not properly addressed, facilities with high hazard potential will experience the same issues repeatedly. As a result, they question what they may have overlooked during the engineering process. After conducting numerous incident investigations and work-related discussions, it is commonly found that only two or three key factors are responsible for most issues.

'If we fix those issues in the system, we will fix not only that for the future but probably all over the place. It was just a problem that we did not see,' Barnaby explains.

Determining what has a high potential for learning within the organization can be a challenging task. Despite attempts to create a process for identifying high learning potential, it became apparent that it's challenging to determine the appropriate questions to ask. As a result, this approach would entail going through 150 questions. Instead, they have chosen to hold a meeting or discussion with those who are best positioned to offer insights. Based on their experience, designing a process for this purpose is a complex undertaking.

The most noticeable difference since they started working with high learning potential principles is the mindset of the employees in the organization. It is the realization that people who make mistakes are not about people. They are likely just trying to deal with more difficult circumstances. Phrases like 'people are the problem' have primarily disappeared, especially from senior management.

Secondly, they have realized that workarounds are standard operating disciplines, and people always have workarounds. The CEO became part of this conversation as he also talked about these concepts. Previously, when senior management discussed safety, it was about how many injuries were reported. Today, they talk about the strengths of their barriers.

'We are not there yet to where we want to be, but we are on that journey: From seeing people as a group who, if they would just comply with what we tell them, will be okay, to becoming an organization that is subordinate to the people on the front line who have the answers,' Barnaby continues.

Talking to people, producing 3-minute videos for people to watch, the messaging from upper management, and how it is conveyed to the work floor have led to everyone being on board with what they are doing. These efforts are in conjunction with the sole old-school crunching of incident statistics.

CREATING LEARNING POTENTIAL THROUGH LEADERSHIP

Companies may be burdened with a tendency to place blame on individuals, which has become ingrained in their culture over time. It can take significant, consistent communication for employees to realize that the company is genuinely committed to change. Unfortunately, the broader societal norms may not always align with the organization's new message. The Integrated Energy Company is exploring the possibility of including an assessment of a candidate's attitude during the selection process for leadership positions. By prioritizing safety, the company hopes to select individuals who can help convey this message effectively to others. Ultimately, having senior leaders who understand this necessity will enable the organization to choose the right people for critical leadership roles.

'The top-level effort we have in the company, which has guided us, is our safety leadership principles,' Barnaby says.

The company has established core values, such as respect, inclusion, and 'one team,' and safety is one of them. Safety leadership principles are employed to define the crucial importance of safety. In the past, safety messages often emphasized the need to 'follow the rules,' 'complete what you begin,' and 'hold accountable those who intentionally violate the rules.' However, these messages failed to encourage individuals to speak up and report unsafe conditions; people kept thinking, 'I would rather just shut up and do my job.'

The organization has the human performance team involved and is leading in creating safety leadership principles. Conflicting efforts to their safety leadership principles get blocked within the company. It works on two fronts: it not only educates people, but it also harmonizes efforts. When they had a reorganization, there was no point in creating different sets of safety leadership principles; they wanted to create one set that applied to the whole organization. Part of that set of principles was to get out there and talk to people, find out what they are doing and what is happening in their world, and understand why mistakes occur in daily operations.

'It's not just headline stuff or one-liners; it guides our actions. In this sense, language can empower or devaluate the message,' Barnaby explains.

SAFETY LEADERSHIP

The IMO (2014) defines leadership as 'a process where one group of individuals influenced by an individual tries to achieve a common goal.' Wu (2007) defined safety leadership as 'the process of interaction between leaders and followers, through which leaders can exert their influence on followers to achieve organizational safety goals under the circumstances of organizational and individual factors.'

Scholars (Cooper, 2015) divide leaders into two types: positional and inspirational. Positional leaders use their power to give orders that must be followed by lower-ranked employees. Inspirational leaders motivate others with their passion and enthusiasm for a shared goal. Leaders typically use one of three leadership styles: transformational, transactional, or servant (Zenger et al., 2009).

Transformational leaders work to change the company culture to achieve their goals. They challenge existing assumptions and motivate employees to exceed their limits. This type of leadership involves four behaviors: idealized influence, inspirational motivation, intellectual stimulation, and individualized consideration. Idealized influence involves setting high moral standards and being a role model for subordinates. This builds employees' trust, admiration, and loyalty (Bass, 1985; Kapp, 2010; Hoffmeister et al., 2014; Pilbeam et al., 2016; Avolio et al., 1999; Bass et al., 2003).

In active transactional leadership, there is a set of standards for performance and the leader can punish employees if they don't follow them. The leader manages closely and corrects any mistakes or deviations from the standards to keep everyone on track toward the goal.

Transactional leadership is different from transformational leadership, and there are three types: constructive, corrective, and laissez-faire. In constructive leadership, rewards are given to encourage good performance. In corrective leadership, mistakes are corrected. In laissez-faire leadership, the leader doesn't usually control performance, only in emergencies. Good communication is important in constructive leadership (Zohar, 2002a, b; Podsakoff et al., 1982).

Servant leaders help their subordinates do their jobs well and keep the company's culture and goals on track. They get to know their subordinates, communicate openly, support their work, and recognize their potential. They also attend safety meetings, support employee ideas, and ensure safety measures are effective (Cooper, 2015).

LANGUAGE AND SHAPING STORIES FROM INCIDENTS

The Integrated Energy Company began their journey of learning from daily work by focusing on incident investigations that required safety leadership thinking. When senior leaders are presented with metrics that stem from incident investigations, indicating that someone made a mistake, neglected to do something, or became complacent, it shapes their mindset.

Barnaby explains: 'We had to change the capability of our investigation team completely. As a result, leaders are now getting the story: *"This is what happened because we did a management change within that team and left them with two rather than four people,"* for example.'

> There were 350 incidents in which the company was trying to get rid of the person who was blamed, and in 320 instances, we were able to show that the person was set up for failure.
>
> – *Barnaby Annan*

The change in language usage made the senior leaders realize they would have done the same in that situation. Management started to see it was about something other than specific people, but people were sometimes set up for failure without realizing it. The change of the stories from the incident investigation gave them a platform to develop a human performance program. This led to where the Integrated Energy Company stands now by asking questions such as *'How can we look at normal work?'* Management had to start at the end: the incident investigations changed the narrative to get people in the right mindset.

That does not mean that some legacy thinking about incidents is wholly gone. Whenever there is a Health, Safety & Environment (HSE) incident, some people still grab back to 'whodunnit,' and that is impossible to change, probably until these people flow out of the organization.

'We must not forget that we cannot just read theoretic books, dump that theory onto an organization, and expect change. The people from the organization need to go on that journey as well. Of course, embarking on that journey will not be perfect, but if you take a few steps towards improvement, you will leave the organization in a better place than where it started,' Barnaby says.

'DON'T LOSE SITUATIONAL AWARENESS!'

Barnaby continues, 'I kept thinking, *"Please don't go around telling people to be situationally aware."*'

Nevertheless, that was what management wanted to do. However, there are limited options in telling people to prevent them from doing that. The organization started working with the initiators of the idea of situational awareness and worked toward the concept. Instead of telling personnel to be more aware, they shifted toward training people to recognize those elements likely to cause certain situations in which they will likely lose situational awareness. Essentially, they are training people to recognize performance-shaping factors, for example, repetitive tasks that will cause them to lose focus. Ultimately, people are then given strategies for managing those situations. Barnaby recognizes this approach is imperfect, but it is better than telling people to be more situationally aware.

SITUATIONAL AWARENESS

Situational awareness is a cognitive process that involves the perception and comprehension of relevant information in one's environment, the projection of future events based on that information, and the ability to make informed decisions and take appropriate actions (Endsley, 2017).

Although human performance principles and performance-shaping factors share a similar, they each have distinct focus areas.

Human performance principles refer to attention management, skill acquisition, decision-making, teamwork, and stress management by looking at the cognitive and behavioral processes underlying human performance.

Performance-shaping factors, on the other hand, refer to environmental, organizational, and individual factors that can impact human performance. Factors may include the physical environment, equipment, and personal characteristics of the individual or team involved.

PERFORMANCE-SHAPING FACTORS

Performance-shaping factors (PSFs) describe the influence of given contexts on human performance and are used to quantify human error probabilities. Examples of PSFs are task complexity, procedures, ergonomics, safety climate, training or experience, fatigue, time pressure, and stress (Liu, et al., 2021).

REIMAGINING SAFETY: THROUGH THE LENS OF TRUST

In addition to the above-described efforts, Barnaby shares that the Integrated Energy Company conducts an annual employee survey about trust in immediate management. Their data science team found an 'overwhelmingly strong correlation' between process safety and trust. Teams with more trust in their leaders also report a higher perception of process safety. Barnaby attributes safety leadership as essential in building a safe climate and trust, which might lead to better safety performance.

In theory, safety and trust seem to be closely related concepts:

1. Trust promotes safety: Mayer et al. (1995) found that trust between coworkers was a critical factor in promoting safety behavior in the workplace. When employees trust each other, they are more likely to communicate effectively, share information, and work collaboratively to prevent accidents.
2. Safety promotes trust: A safe working environment can help build employee trust. For example, Zohar (1980) found that when employees perceived their workplace as safe and supportive of their well-being, they were more likely to develop positive attitudes toward their colleagues and supervisors.
3. Trust can mitigate risk-taking behavior: In high-risk industries such as aviation and healthcare, risk-taking behavior can seriously affect safety

outcomes. Salas et al. (2006) found that when team members trusted each other's expertise and judgment, they were less likely to engage in risky behaviors during critical tasks.

4. Trust can improve reporting of safety incidents: Organizations must have accurate information about incidents to address safety issues effectively. However, employees may be reluctant to report incidents if they fear punishment or retaliation. Brimhall et al. (2023) found that when employees trusted their organization's commitment to safety and felt safe in reporting incidents, they were more likely to report medical errors.

THE ONLY CONSTANT IN LIFE IS CHANGE

The stories of professionals throughout this chapter highlight that change in how we think about safety and how we practice safety is not a straight line toward ultimately safe. The key is to keep looking for what works for the organization. We must acknowledge that nobody holds the ultimate truth about safety. What works for another might not work for someone else, and vice versa. We cannot stop asking genuinely curious questions about what and why of someone's intentions – whether it be an operational worker or a senior manager. As we have seen in Jaap's story, creative and out-of-the-box actions can create ripple effects throughout the company. As the QHSE manager explained, making changes requires strength and courage, especially if the professional culture is opposed. As Barnaby explained, efforts to improve safety might be a prolonged process, require a radical change in mindset, and need to have parallel efforts. All the while, how we talk to ourselves and others might be the starting point for positive changes. As professionals in safety, we need to remain open-minded to these stories.

CONCLUSION

The first chapter of this book has established a compelling case for the integral role of trust and open communication in enhancing safety within the workplace. The real-life stories have shown that fostering a culture of trust can significantly mitigate risk-taking behavior, improve incident reporting, and overall, boost safety performance. The relationship between trust and safety is undeniable. Trust fosters an environment that encourages open communication, thereby promoting safety-conscious behaviors and reducing risks. The process of improving safety is continuous and requires an openness to innovation and the courage to challenge conventional norms. Making a significant impact on safety often requires thinking outside the box and being willing to take calculated risks. Organizations can move toward turning risks into resilience, ultimately becoming safer, more trusting, and innovative workplaces by considering the following:

1. **Encourage open communication:** Create an environment where employees feel safe to speak up about potential safety issues.
2. **Promote trust-building activities:** Foster trust within the team and encourage transparency. This could involve sharing company updates, celebrating team achievements, and addressing challenges collectively.

3. **Continuously innovate safety practices:** Regularly review and update safety practices to ensure they are effective and fit for purpose. Be open to new ideas and encourage employees to contribute their suggestions for improvement.

REFERENCES

Avolio, B. J., Bass, B. M., & Jung, D. I. (1999). Re-examining the components of transformational and transactional leadership using the Multifactor Leadership Questionnaire. *Journal of Occupational and Organizational Psychology*, 72, 441–462.

Bass, B. M. (1985). Leadership: Good, better, best. *Organizational Dynamics*, 13, 26–40.

Bass, B. M., Avolio, B. J., Jung, D. I., & Berson, Y. (2003). Predicting unit performance by assessing transformational and transactional leadership. *Journal of Applied Psychology*, 88, 207–218.

Brimhall, K., Tsai, C., Eckardt, R., Dionne, S., Yang, B., & Sharp, A. (2023). The effects of leadership for self-worth, inclusion, trust, and psychological safety on medical error reporting. *Health Care Management Review*, 48(2), 120–129. doi: 10.1097/HMR.0000000000000358.

Cooper, D. (2015). Effective safety leadership: Understanding types & styles that improve safety performance. *Professional Safety*, 60, 49–53.

Endsley, M. R. (2017). From here to autonomy: Lessons learned from human–automation research. *Human Factors*, 59(1), 5–27. doi: 10.1177/0018720816681350.

Hoffmeister, K., Gibbons, A. M., Johnson, S. K., Cigularov, K. P., Chen, P. Y., & Rosecrance, J. C. (2014). The differential effects of transformational leadership facets on employee safety. *Safety Science*, 62, 68–78.

International Labor Organization (ILO). (2020). Quick Guide on Sources and Uses of Statistics on Occupational Safety and Health. Retrieved from: https://www.ilo.org/wcmsp5/groups/public/---dgreports/---stat/documents/publication/wcms_759401.pdf

International Maritime Organization (IMO). (2014). *Model course 1.39: Leadership and teamwork*. London, UK: International Maritime Organization.

Kapp, E. A. (2012). The influence of supervisor leadership practices and perceived group safety climate on employee safety performance. *Safety Science*, 50, 1119–1124.

Liu, J., Zou, Y., Wang, W., Zhang, L., Liu, X., Ding, Q., Qin, Z., & Čepin, M. (2021). Analysis of dependencies among performance shaping factors in human reliability analysis based on a system dynamics approach. *Reliability Engineering & System Safety*, 215, 107890.

Mayer, R. C., Davis, J. H., & Schoorman, F. D. (1995). An integrative model of organizational trust. *The Academy of Management Review*, 20(3), 709. doi: 10.2307/258792.

Pilbeam, C., Doherty, N., Davidson, R., & Denyer, D. (2016). Safety leadership practices for organizational safety compliance: Developing a research agenda from a review of the literature. *Safety Science*, 86, 110–121.

Podsakoff, P. M., Todor, W. D., & Skov, R. (1982). Effects of leader contingent and noncontingent reward and punishment behaviors on subordinate performance and satisfaction. *Academy of Management Journal*, 25, 810–821.

Salas, E., Rosen, M. A., Burke, C. S., Goodwin, G. F., & Fiore, S. M. (2006). The making of a dream team: When expert teams do best. In K. A. Ericsson, N. Charness, P. J. Feltovich, & R. R. Hoffman (Eds.), *The Cambridge handbook of expertise and expert performance* (pp. 439–453). New York: Cambridge University Press.

Stanton, N. A., Salmon, P. M., Walker, G. H., Salas, E., & Hancock, P. A. (2017). State-of-science: Situation awareness in individuals, teams and systems. *Ergonomics*, 60(4), 449–466. doi: 10.1080/00140139.2017.1278796.

Wu, T. C., Chen, C. H., & Li, C. C. (2007). Correlation among safety leadership, safety climate and safety performance. *Journal of Loss Prevention in the Process Industries*, 6, 261–272.

Zenger, J. H., Folkman, J. R., & Edinger, S. K. (2009). *The inspiring leader: Unlocking the secrets of how extraordinary leaders motivate.* New York: McGraw-Hill.

Zohar, D. (1980). Safety climate in industrial organizations: Theoretical and applied implications. *Journal of Applied Psychology*, 65(1), 96–102.

Zohar, D. (2002a). The effects of leadership dimensions, safety climate, and assigned priorities on minor injuries in work groups. *Journal of Organizational Behavior*, 23, 75–92.

Zohar, D. (2002b). Modifying supervisory practices to improve subunit safety: A leadership-based intervention model. *Journal of Applied Psychology*, 87, 156–163.

2 The Good, the Bad, and the Ugly
The Evolution of Safety

It is not the strongest of the species that survive, nor the most intelligent, but the one most responsive to change.

– Charles Darwin

INTRODUCTION

People's view of safety, risks, and hazards has changed over time. In the past, accidents were often attributed to divine intervention, and those responsible were punished for their mistakes. The field of safety is constantly evolving to meet society's changing needs and concerns.

Unfortunately, history has shown that incidents and accidents often precede the need or the will to change. Moreover, modern safety thinking sometimes seems to regress. In certain incidents, the good, the bad, and the ugly all come to light, and the evolution of safety can be encapsulated in a single event.

THE UGLY: THE DELTA INCIDENT

Airports are bustling hubs of activity, with operations occurring both on the ground and in the air. This is attributed to the various users, not just aircraft but also ground handlers, ground equipment, and airport vehicles. Managing all this traffic and keeping it separate in the Maneuvering Area falls primarily under the responsibility of air traffic controllers. Interestingly, incidents are more prone to happen on the ground than in the air. Up to 80% of reported incidents take place on the ground.

Like road traffic, and within the Maneuvering Area, airport ground traffic has rules and procedures that must be followed. Occurrences may happen if procedures and air traffic controllers' (ATC) instructions are not correctly followed. An example might be taking the wrong turn or not giving way to other vehicles or aircraft. While most occurrences do not result in any damage or loss, controllers still report them to the safety department for trend tracking and investigation purposes. These reports provide valuable insight into the operation, and the safety department relies on controllers to provide them. On one particular occurrence, a serious accident was narrowly avoided.

On December 10, 1998, at Amsterdam Schiphol Airport, a Delta Air Lines Boeing 767 received a takeoff clearance from the air traffic control for departure from runway 24.

DOI: 10.1201/9781003383109-2

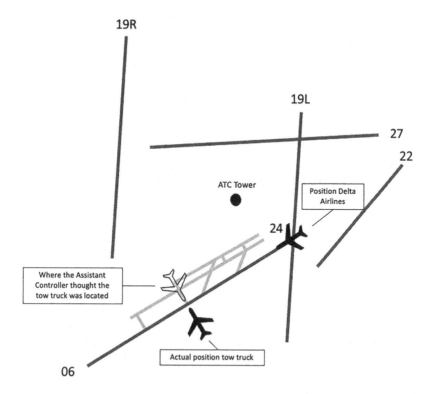

FIGURE 2.1 Runway layout Schiphol Airport in 1998 with the position of the tower, the tow truck and Delta Airlines aircraft, and where the Assistant Controller thought the tow truck was positioned.

Further down the runway, however, there was a Boeing 747 being towed from one side of the runway to the other by a tow truck, crossing in front of the departing aircraft. This aircraft had already received clearance to cross the runway from the Assistant Controller. Figure 2.1 shows a schematic representation of the situation.

At the time of the incident, low-visibility procedures were in force. Because of the limited visibility, the Delta crew saw the towed aircraft after they started their takeoff roll. The pilots of the Boeing 767 aborted the takeoff when they observed the crossing of the Boeing 747 crossing in front of them. The incident was reported to the Dutch Safety Board (DSB) and the Dutch Air Traffic Control (Luchtverkeersleiding Nederland, LVNL).

INCIDENT DEVELOPMENT

The Assistant Controller reported for duty, unaware that a newly constructed tunnel had been built beneath runway 24 to facilitate the passage of small vehicles, like tow trucks, to the other side of the runway. When the tow truck with the Boeing 747 aircraft behind it requested to cross the runway, the Assistant Controller assumed they were crossing from Schiphol Centre to the East of the airport, as they were not visible from the tower due to the weather. The working position of the Assistant was not equipped with a ground radar that could be consulted. However, the crossing was actually taking place from the East to Schiphol Centre. The Assistant Controller forwards the request to cross to the Runway Controller, which grants this permission. Then, the Assistant Controller instructs the tow truck to cross the runway and deactivates the set of lights that protect the runway crossing.

However, the tow truck driver reported that the stop bar lights are still illuminated (which is true as he is standing on the other side of the runway). The Assistant Controller and the Runway Controller discussed the issue. Recently, a new panel with light controls was installed; perhaps, it wasn't working as it should. The Assistant Controller pushed all the buttons, after which the stop bar light is deactivated and the light extinguished, and the tow truck driver is now free to cross the runway.

INCIDENT ANALYSIS

Various factors influenced the development of the incident development. The Assistant Controller – tasked with issuing a crossing clearance under the responsibility of the Air Traffic Controller – did not have access to a ground radar picture display, causing uncertainty about the tow's actual position. Furthermore, the Runway Controller – responsible for the runway – was a Trainee under the supervision of the Instructor – who was also on duty as a Tower Supervisor. The labeling and functionality of the newly added panel controlling the stop bars at runway intersections were not immediately clear. Despite initial signs indicating not everything was in order, the Trainee Controller cleared the Delta Airlines aircraft for takeoff, and the Instructor of the Air Traffic Controller Trainee was simultaneously performing duties as Supervisor duties in the tower.

After analysis, LVNL's report recommended 23 corrective actions to address systemic issues in system design, layout, staffing, instructing, communication, and handovers. The DSB report listed causal factors being low-visibility weather conditions, inadequate information during radio communication with the tow combination, misinterpretation of the tow's position and movement, insufficient teamwork and supervision, and bad confirmation of runway clearance before takeoff. The DSB issued similar recommendations to rectify systemic deficiencies in the organization of ATC The Netherlands (LVNL) and Schiphol Airport.

THE BAD: THE CRIMINAL PROSECUTION

Unfortunately, the crew of Delta Airlines decided to take LVNL and the controllers to court. The persons prosecuted were the Instructor/Supervisor, the Trainee Controller, and the Assistant Controller.

Under Dutch law, it is 'forbidden to provide air traffic control in a dangerous manner, or in a manner that could be dangerous to persons or properties.' Two years after the incident, the aviation prosecutor formally charged the Instructor/Supervisor, the Trainee Controller, and the Assistant Controller with breaking this part of the law.

Under Dutch law, it is 'forbidden to provide air traffic control in a dangerous manner, or in a manner that could be dangerous to persons or properties.'

– Article 5.3

The first criminal court case was held in August 2001. The judge ruled that the Assistant Controller was not guilty but that both the Trainee and the Instructor/ Supervisor were. As a result, they were sentenced to a fine of about 450 US dollars or 20 days in jail.

The decision was appealed by both the Trainee Controller and the Instructor/ Supervisor, while the prosecutor appealed against the Assistant Controller's acquittal. Finally, in September 2002, a panel of three judges presided over the case in a higher court. As part of the legal process, they, along with the prosecutor and their legal team, were taken to the airport's tower ('the scene of the crime'), where safety-critical work was carried out to gain a firsthand understanding of the crime scene.

The only admissible defense against this determination is being devoid of all blame which could only succeed if a Controller was off-duty and therefore not in the tower.

– Skybrary.aero (1998)

The court found all three accused guilty, but no sentences were imposed, as the case was treated as an 'infringement of the law' and not an 'offense against the law.' In Dutch law, an infringement is considered as guilt; blame is assumed to be present and doesn't need to be proven. The only defense against this determination is being devoid of all blame, which can only be granted if a Controller is off-duty and not in the tower. The decision also eliminates any appeal option, as no conviction of an offense or punishment was given.

The court acknowledged that the control tower facilities for preventing such incidents were suboptimal, as shown by improvements implemented after the incident. The court recognized the prosecution of the three controllers for their professional duties' infringement, which deeply affected their personal and professional lives.

The court took into consideration the prosecutor's indication that this case was a legal 'test case.' The judgment stated that none of the defendants had a criminal record, and there were no indications that they had ever failed in their responsible functions.

THE UGLY – AGAIN – THE AFTERMATH

The effects of the criminal prosecution went beyond the profound impact on the controllers' lives. The results were visible throughout the entire organization, the air traffic control community, and the development of aviation safety.

Once the dust had settled from the prosecution, the controllers experienced a disturbance in their professional relationships and a lack of protection to perform their duties during, from their organization, and under Dutch law. As a result of the experienced psychological unsafety, the operational staff became reluctant to cooperate in incident investigations. The just culture was shattered, and their willingness to

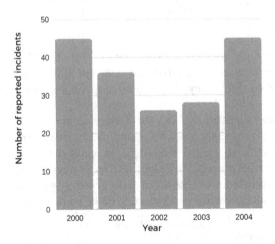

FIGURE 2.2 Number of reported ATC incidents by air traffic controllers after the Delta incident (courtesy of Job Brüggen).

report decreased significantly. To this day, LVNL is one of the only Air Navigation Service Providers that has measured the drop in the voluntary reports in ATC after the prosecution, as shown in Figure 2.2. Fewer reports mean less opportunity to learn from incidents and delay further development of the safety management system (SMS) and the safety information flow within the organization. Eventually, reporting was restored to pre-incident levels at LVNL, but there were years of lost potential organizational learning.

JUST CULTURE

Just culture is an atmosphere that promotes trust, emphasizing the importance of learning, accountability, and improvement rather than blame and punishment. In a just culture, individuals are encouraged to report errors and near misses without fear of retribution, fostering a culture of open communication and transparency (Reason, 2000; Marx, 2013).

The core principles of just culture are as follows:

1. Accountability: Individuals are held responsible for their actions and decisions, but the focus is on understanding the underlying factors that contributed to the error rather than solely blaming the individual.
2. Learning: Mistakes and adverse events are viewed as opportunities for learning and improvement. Organizations strive to identify the root causes of errors and implement system-level changes to prevent them from recurring.
3. Trust: Transparency and mutual respect are crucial in a just culture. Individuals feel safe to report errors and share information without fearing negative consequences.

THE GOOD: LESSONS LEARNED

The Delta Airlines incident allowed the prosecutor to open the closed stronghold of Schiphol's air traffic control operation. After the incident, LVNL took a much more proactive approach in reporting about their operation, their incidents, and their plan to improve toward the aviation authorities.

In the Netherlands, any reporter of an occurrence has been legally protected since 2006. In short, this protection means that the reporter of an incident does not have to fear prosecution (criminal, administrative, or civil) if there is no sign of willful misconduct or gross negligence. This protection is also included in European Regulation 376/2014.

Someone who witnessed the aftermath of the prosecution firsthand is Job Brüggen – Safety Officer at LVNL.

'I feel strongly both ways. The Tower crew certainly did some things that could have been done otherwise professionally. Nevertheless, who is to blame? The organization, of course. The organization has selected them, trained them, allowed their work with *their* systems and materials with their own designed procedures, in a culture that *they* nourish,' Job says.

Job explains he sees this incident as beautiful, in a way; there was no damage, no injuries, and no fatalities, but the organization has learned so much from it. Afterward, it was the starting point of many changes within the organization and the sector.

LEARNING THAT TRANSCENDS ORGANIZATIONAL AND NATIONAL CULTURE

Since the Delta incident, legislation has changed on both a national and a European level. The European Commission established a directive numbered 2003/42. In this directive, operators' reports are exempt from the Government Information (Public Access) Act, which provides for the active and passive disclosure of documents. Instead, a summary of the reports is sent to the ABL (*Analysebureau Luchtvaartvoorvallen*, Aviation Occurrence Analysis Agency), which is also exempt from public access under the Government Information Act. One of the reasons for this change is to enable reporters to freely report incidents, even if they are a causal factor during the development of an incident, without the risk of 'writing their own fines.'

ABL was established due to a 'fundamental distrust between the government and sector stakeholders' (aviation parties): It is tasked with properly informing the government and the Transportation Inspectorate and to perform trend analysis on the reported events.

Under the Balkenende administration, the Dutch government was reluctant to establish new laws and rules. Still, they did so anyway: The Aviation Act changed into what is now known as Regulation (EU) No 376/2014, which states that reporters of occurrences cannot be prosecuted based on a report (unless there was willful misconduct or gross negligence). Again, the Dutch were at the forefront of the law and was later included in European law.

In the National Aviation Act, 11.25 and 11.26, the public prosecutor is not allowed to sniff around in the desk drawer of the safety manager when the organization has a certified SMS in place.

SAFETY MANAGEMENT SYSTEM

An SMS is a systematic approach that organizations implement to proactively manage and improve safety. It involves a set of policies, procedures, and practices designed to identify and mitigate risks, prevent accidents, and ensure the well-being of employees and the public. An SMS typically consists of four key components:

1. Safety policy: The organization establishes a clear safety policy that outlines its commitment to safety, defines roles and responsibilities, and sets safety objectives and targets.
2. Risk management: A systematic process is employed to identify hazards, assess risks, and implement control measures to mitigate those risks. This involves conducting risk assessments, analyzing potential consequences, and implementing appropriate risk controls.
3. Safety assurance: Ongoing monitoring and evaluation are conducted to ensure the effectiveness of safety measures. This includes regular safety audits, inspections, and performance monitoring to identify gaps or deficiencies and take corrective actions.
4. Safety promotion: Organizations foster a safety culture by promoting awareness, training, and communication. It involves providing education and training programs, encouraging reporting of safety concerns, and facilitating open communication channels.

(ICAO, 2013; ISO, 2018)

Despite the positive changes in aviation laws, 'Article 5.3' still hovers above the new laws. This article states that it is 'prohibited to participate in air traffic or to provide air traffic control services in such a way that persons or property are or may be endangered,' which of course can only be guaranteed if no one is flying and no is providing air traffic control. However, there has been an agreement between the government and the aviation sector parties that this article will not be brought up immediately, as was done after the Delta incident. The article gives the public prosecutor some final leverage but is not intended as the starting point after investigating a safety event. It will only be used as a 'last resort.'

PROTECTING THE REPORTERS

The relationship between the public prosecutor and the aviation sector has significantly improved. However, in other industries (such as healthcare) and even other European countries, the prosecution of operational personnel is still taking place today. For example, in 2013, a serious incident occurred in Swiss airspace where two aircraft were at a dangerous convergence, but no collision occurred. The serious incident is attributable to the fact that the crew of one aircraft initiated a climb based on a clearance issued to another aircraft from the same airline. As a result, the airline Captain and the Air Traffic Controller were prosecuted, and both were found guilty in 2018. The Controller tried to appeal, but the Supreme Court rejected it. Investigations were conducted, but no criminal charges were filed.

REPORTING: WHY ONE WOULD OR WOULDN'T

Although reporting occurrences sounds logical and desirable, why wouldn't anyone report safety issues? Bart Runderkamp, a Captain at a European airline, describes it as follows:

> When safety is compromised, we're required to write a report, for example, if a runway incursion occurred, loss of communication with ATC, an engine failure, a bird strike, ground handling not following procedures, or discrepancies in the papers (for example, the load sheet). The safety department then forwards the anonymized reports to the relevant agencies.

However, it's not uncommon for a Captain to heave a sigh when something that needs to be reported occurs. Some Captains don't report always, and usually, the demographic is that these Captains are more experienced (not to say 'older'), flying for over 20 or 30 years. In their experience, throughout the years, they have had to do increasingly more. In their logic, they've been flying safely for a long time and don't see the added value of reporting.

This doesn't apply to all; most pilots do report. Bart always writes his reports with as many details as possible, not just to be complete but also to avoid being called later by the safety department asking for more information. To Bart, it's not just about contributing to increasing safety but also avoiding later troubles. Finally, reporting is one of the few things they can do to improve safety – other than flying, complying with rules and procedures, getting trained, etc.

Reporting via a report for data gathering and analysis is of immense value to the organization. However, sometimes, 'reporting' happens – or should occur – during live operations, also known as 'speaking up.'

THE NEED FOR SPEAKING UP VERSUS THE NEED TO NOT LOOK STUPID

Reporting – whether it's a report or in real life – requires more than 'just saying it.' It requires psychological safety.

'A pilot's worst fear,' Stephen Walsh, Managing Director, Interaction Trainers, explains, 'is to look stupid. How do you give someone the confidence and the ability, as a First Officer, in the middle of the night, in a situation in which they are thinking, "*I hope the Captain knows what he or she is doing*," and keeping their mouth shut, instead of taking action?'

Stephen has spent nearly three decades training pilots in crew resource management (CRM), a soft skills training to improve communication and optimal use of available resources.

'What pilots tend to suffer from is something called "loss aversion." It's a state in which you're afraid to admit that you don't know where you are to the other person but will not say anything about it to avoid being seen as incapable of doing your job.

It's the fear that your colleague will tell others, and it might end up with a senior who might question your abilities and lose your livelihood.

Stephen believes pilots and other professionals can be trained to acknowledge their vulnerability and overcome "loss aversion"; "What happens is that other colleagues will come up to you and shake your hand because you're a safe pilot (not afraid to admit when they are lost, wrong or do not understand something) and they want to fly with you."

People tend to emphasize what they can lose rather than what they can win. People need the safety that if they speak up, they will be rewarded by positive social reinforcement, a "well done, thank you." And that's how training should be used: to incentivize positive reinforcements on behaviors you want to see.'

It's not about being nice to each other; it's about being professionally effective.

– Stephen Walsh

CRM began to evolve as a particular result of the Tenerife accident in 1977. During this accident, two aircraft collided on the runway after a KLM Boeing 747 took off while another Pan Am Boeing 747 was still taxiing out on the runway. The causal factors of the Tenerife accident are often cited as miscommunication and poor visibility due to fog. The KLM Captain, believing he had received clearance to take off, began rolling the aircraft down the runway while the Pan Am aircraft was still taxiing to the terminal. The KLM crew missed several opportunities to verify their takeoff clearance with ATC, and the Pan Am crew was unaware of the KLM aircraft's location due to the fog. However, the Cockpit Voice Recorder revealed that there was also miscommunication inside the cockpit of the KLM aircraft. The Flight Engineer expressed concern about this action multiple times, but the Captain dismissed these concerns. The Captain began the takeoff roll without confirming the runway was clear. When he saw the Pan Am aircraft, it was too late. The KLM aircraft collided with the Pan Am aircraft, resulting in the death of 583 people.

Accidents of this scale can never be attributed to one 'root' cause as several factors were at play. The cause of the miscommunication has been attributed to multiple factors, including the Captain's 'macho and anti-authority attitudes,' as some would describe it. Additionally, the KLM flight crew was under time pressure because of flight time limitations, and the airport was experiencing heavy congestion and radio interference, which blocked out several transmissions made by ATC and flight crews.

In the aftermath of the accident, KLM asked Interaction Trainers to create a course for their Captains to address the identified issues. The first courses were titled 'The Captain's Course.' Later, it became CRM. After the Captains began to be trained on CRM topics, KLM Flight Engineers said, '*Hang on a second, it was our Flight Engineer who's questioning the Captain dismissed; we want this training too!*' KLM then decided that all flight crew should have the training (Air International Magazine, 2019).

ASK THE TRAINEES RATHER THAN TELLING THEM WHAT TO DO

In Stephen's experience, the most effective instructors are the instructors who use a *question-based method of teaching*. It's not their job to tell somebody things they already know. It's an instructor's job to ask questions about what they know and don't know and help them find out what they don't know *yet*. In the flight crew's case, how can an instructor overcome their worst fear? And their worst fear is to look stupid.

Another example Stephen gives: During simulator training, he would ask First Officers, '*If you see a situation unraveling unsafely, what would you do?*' and their responses would always be, '*I would take control.*' But during the simulation sessions, Stephen would simulate the Captain – as pilot flying – having a brain seizure during the final approach. Not being dead, no, they would have a brain seizure, meaning they would not move their head nor respond in any way, having the flight controls still in their hands. At some point during the simulation session, Stephen would tap the Captain on the shoulder, and they would fake the brain seizure while the First Officer was checking the instruments and making the callouts. '500' (altitude from runway in feet, red.) would be the altitude the First Officer would call out, and the Captain would not respond.

The reason airline procedures have those calls is that if a response is withheld, the other pilot must act immediately. But the First Officers didn't do that; up to three or four times, they would hesitate. Both the Captain and First Officer were briefed that this would happen beforehand. 'Why didn't you immediately take control?' Stephen would ask them. They all come up with the same answer: the thought of

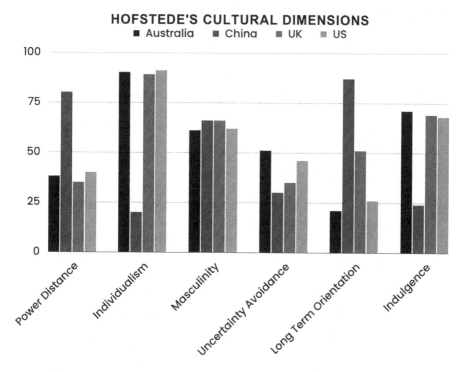

FIGURE 2.3 Scores on Hofstede's Cultural Dimensions by Australia, China, the UK, and the US.

taking control from a living, breathing Captain with eyes wide open was terrifying. Unfortunately, hesitancy can cause serious adverse effects.

> The thought of taking control from a living, breathing Captain with eyes wide open was terrifying. But hesitancy can cause serious adverse effects.
>
> – *Stephen Walsh*

Although it sounds easy to shout, 'people should speak up,' it's not as simple as it seems. Some behaviors people show are so ingrained within a culture – whether organizational or even national – that it's unrealistic to expect them to break the pattern.

One theory that emphasizes cultural influences is Hofstede's culture index. It shows that some behaviors are more prominent in some parts of the world than others, as seen in Figure 2.3.

In the above figure, it becomes clear that in some countries, Power Distance is scored as high, for example, in China. However, in Chinese culture, the tolerance for uncertainty is relatively lower. It's important to note that there is no right or wrong; it's just a map of cultural preferences, values, and acceptance.

Hofstede's cultural dimensions theory is a framework for cross-cultural communication that describes the effects of a society's culture on the values of its members and how these values relate to behavior. It consists of six dimensions that can be used to describe different aspects of culture:

- Power distance: The extent to which the less powerful members of a society accept and expect that power is distributed unequally.
- Individualism vs. collectivism: The degree to which individuals value individualism or collectivism.
- Masculinity vs. femininity: The distribution of emotional roles between the genders.
- Uncertainty avoidance: The degree to which the members of a society feel uncomfortable with uncertainty and ambiguity.
- Long-term vs. short-term orientation: The degree to which society embraces, or does not embrace, long-term devotion to traditional, forward-thinking, or pragmatic values.
- Indulgence vs. restraint: The degree to which people attempt to control their desires and impulses.

An organization can make all the 'Be safe' or 'Speak up' posters in the world, but it won't make a difference if the employees are not empowered in every possible way. It's the job of the training department to make people capable of doing so, and according to Stephen, 'it's the most important job in the world.'

A PARALLEL DEVELOPMENT: PSYCHOLOGICAL SAFETY

In recent years, the concept of psychological safety has taken a prominent place on a societal level.

PSYCHOLOGICAL SAFETY

Psychological safety is the belief that one can express their opinion or make a mistake without fear of negative consequences. Initially, organizational psychologist Amy Edmondson introduced the concept of psychological safety in 1999 describing that teams with a high level of psychological safety displayed a willingness to learn from their mistakes and were more effective in problem-solving than teams with low levels of psychological safety. Since then, psychological safety has gained traction in various fields, from a relatively unknown concept to a widely recognized and important aspect of teamwork and organizational success (Edmondson, 1999).

Both just culture and psychological safety share an emphasis on openness, trust, and learning. They both require a non-punitive response to mistakes and encourage employees to voice their concerns. However, there are differences between the two concepts. Just culture primarily focuses on how *organizations* respond to errors and adverse events, emphasizing system improvements and accountability. Psychological safety encompasses a broader scope and includes aspects of *interpersonal interactions* in the workplace, not just those related to errors or safety incidents.

Just culture can be seen as a component of psychological safety or vice versa. A just culture contributes to psychological safety by ensuring employees feel safe reporting errors or safety concerns. On the other hand, psychological safety goes beyond error reporting and covers all forms of employee expression and interaction.

Organizations striving for a just culture and psychological safety often implement open-door policies, anonymous incident reporting, regular feedback sessions, and training on open communication and empathy.

Studies have shown that psychological safety is positively associated with employee engagement, creativity, and innovation (Frazier et al., 2017; Newman et al., 2017). Conversely, when psychological safety is lacking, employees may be less likely to speak up or take risks, leading to missed opportunities for improvement and growth (Edmondson & Lei, 2014).

It's becoming increasingly clear that psychological safety is no longer a nice-to-have but a necessity. Organizations increasingly recognize leaders' role in creating psychologically safe environments. In aviation, they have already shifted toward European legislation that mandates organizations to implement this framework. For example, EASA requires the implementation of SMSs, and this includes the implementation of just culture.

CONCLUSION

This chapter delved into how safety perspectives and practices have evolved over time. There has been a shift toward fostering learning cultures instead of punitive measures and how this has greatly influenced the development of safety management.

In the past, accidents were often attributed to divine intervention or human error, but now, we recognize the role of systemic factors. By prioritizing learning over blame, organizations can establish effective safety management systems. Building a safety culture and promoting open reporting are essential for managing risks and ensuring business continuity. This chapter stresses the importance of having a just culture and psychological safety in order to encourage open reporting and create a safe working environment.

To advance toward a more resilient and safety-conscious future, organizations should focus on the following:

1. **Promoting a just culture:** Encourage a work environment that focuses on learning from mistakes rather than punishing them. This could involve changing company policies or providing training to managers.
2. **Fostering psychological safety:** Create a safe space where employees feel comfortable voicing their concerns and ideas without fear of negative consequences.
3. **Implementing effective safety management systems:** Regularly review and update safety protocols to encourage open reporting and foster a safety culture.

REFERENCES

Air International Magazine, February 2019, pp. 76–81, author's copy.

B763, Delta Air Lines, Amsterdam Schiphol Netherlands, 1998 (Legal Process – Air Traffic Controller). Retrieved from: https://skybrary.aero/articles/b763-delta-air-lines-amsterdam-schiphol-netherlands-1998-legal-process-air-traffic

Edmondson, A. (1999). Psychological safety and learning behavior in work teams. *Administrative Science Quarterly*, 44(2), 350–383.

Edmondson, A. C., & Lei, Z. (2014). Psychological safety: The history, renaissance, and future of an interpersonal construct. *Annual Review of Organizational Psychology and Organizational Behavior*, 1, 23–43. doi: 10.1146/annurev-orgpsych-031413-091305.

European Aviation Safety Agency. (2020). A guide to implementing a safety management system in accordance with Commission Regulation (EU) No 965/2012, Annex 19.

European Union Aviation Safety Agency. (2021). Commission Regulation (EU) No 965/2012.

Frazier, M. L., Fainshmidt, S., Klinger, R. L., Pezeshkan, A., & Vracheva, V. (2017). Psychological safety: A meta-analytic review and extension. *Personnel Psychology*, 70(1), 113–165. doi: 10.1111/peps.12183.

Hofstede's Cultural dimensions. Retrieved from: https://geerthofstede.com/

International Civil Aviation Organization. (2013). *Safety Management Manual (SMM)* (Doc 9859) (4th ed.). ICAO. https://www.icao.int/safety/SafetyManagement/Documents/Doc9859_4ed_en.pdf

International Civil Aviation Organization (ICAO). (2021). Doc 9859 — Safety Management Manual (SMM).

International Organization for Standardization. (2018). ISO 45001:2018 Occupational health and safety management systems - Requirements with guidance for use. ISO.

Marx, D. (2013). Just culture: A foundation for balanced accountability and patient safety. *Quality and Safety in Health Care*, 13(Suppl. 2), ii5–ii9. doi: 10.1136/qshc.2004.010447.

Newman, A., Donohue, R., & Eva, N. (2017). Psychological safety: A systematic review of the literature. *Human Resource Management Review*, 27(3), 521–535. doi: 10.1016/j.hrmr.2017.01.001.

NTSB Aircraft Accident Report, KLM Royal Dutch Airlines B-747, PH-BUF and Pan American World Airlines, Inc. B-747, N736PA, Tenerife, Canary Islands, Spain, March 27, 1977.

Reason, J. (1990). *Human error*. Cambridge, UK: Cambridge University Press.

Reason, J. (2000). Human error: Models and management. *BMJ*, 320(7237), 768–770. doi: 10.1136/bmj.320.7237.768.

3 Safety Rebels Always Ring Twice

Insanity is doing the same thing over and over and expecting different results.
– Various sources

INTRODUCTION

Referring to the 1934 novel by James M. Cain, *The Postman Always Rings Twice*, the title points to the early days when mail was delivered. When the postman knocked once, the mail had been delivered, and no further action was needed from the recipient. However, if the postman knocked twice, it meant that a telegram was being delivered, which required personal attention from the household. Telegrams were expensive and often brought unpleasant news, so a double knock from the postman was a signal that trouble or bad news was on its way (Flanders, 2003).

I picked this chapter title because it captures the idea of employees having to deliver difficult news, whether operational staff, safety experts, or project managers. It's not unusual – I've personally experienced this – for both management and operational staff to view safety messages as negative news. They may feel like they have to change how they work to be safe (if they are operational staff) or that management is asking for expensive changes to address a growing number of incidents. I have heard people say that their colleagues have 'pulled the safety card to block a project' and safety departments as 'the place where innovation goes to die.'

A positive trend in the last decades is integrating safety into all aspects of an organization's operations, rather than a separate department from other departments, and instead working collaboratively with other teams to identify and mitigate risks. This shift is far more favorable than viewing safety and safety employees as a regulatory requirement or a compliance issue that needs to be managed.

In relatively recent years, I have seen significant developments in safety thinking, with a growing recognition that safety is not just the absence of accidents but also involves proactive risk management. One such development has been the shift toward a more human-centered approach to safety, recognizing that human factors play a crucial role in accidents and injuries.

ORGANIZING WORK IN THE ANGLO-SAXON OR RHINELAND APPROACH

When people visualize how organizations are structured and managed, they usually visualize an organogram, for example, such as in Figure 3.1. An organogram

DOI: 10.1201/9781003383109-3

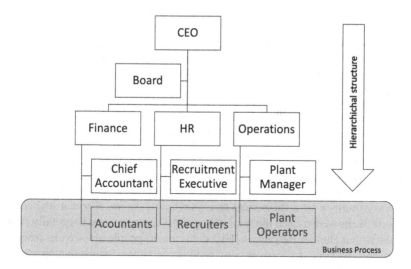

FIGURE 3.1 A typical organogram.

represents a clear hierarchy; there is a clear division between what is considered to be top decision-makers and operational staff; there is a clear leader, people facilitating the leader, people telling other people what to do, and people doing the work.

More and more organizations outsource the bottom layer, the *where-the-work-is-done* departments. The operational divisions may be considered 'difficult' and 'costly'; they require many resources from the organization. Also, it's the place where most of the mistakes happen. Organizations and contracting parties would typically agree on a service-level agreement to ensure everybody knows what can be expected, how much will be paid for this work, and other metrics.

The above hierarchical management structure is a typical example of organizing in an Anglo-Saxon approach. One of the defining features of the Anglo-Saxon work style is a focus on results-driven productivity. This approach emphasizes measurable outcomes and often requires employees to meet or exceed specific performance targets. According to the Organization for Economic Cooperation and Development (OECD, 2004), organizations with an Anglo-Saxon business model tend to focus on 'performance-based pay, hard work, and individual initiative.'

There are four foundations of this way of thinking: Command (the boss is in charge), Control (trust is nice, but having control is better), Communication (*to* the people, not *with* the people), and finally, Intelligence (data is needed to keep everything in check). This way of organizing and management goes hand in hand with measuring a set of key performance indicators (KPIs) and dashboards.

In these types of organizations, it doesn't matter what the background of the person in charge is – whether that's the CEO or management level. A manager could equally be a leader in a hospital as well as a commercial cookie company: knowledge on the subject isn't necessary because they're looking at the same structure with the same type of data.

SAFETY TRAINING ≠ GUARANTEED SAFETY IMPROVEMENT

Jurriaan Cals, Organizational Advisor at Artemas, was trained at the fire service academy in the Netherlands. In his first job, he started working in crisis management. Crisis management is the process of preparing for, managing, and mitigating the impact of unpredictable events that can harm an individual, a company, or a community. It involves identifying potential crises, creating a plan to address them, and executing it when a crisis occurs.

While studying at the fire service academy, Jurriaan noticed that all fire service-related topics were trained at the academy itself. However, topics like communication and leadership skills were outsourced to a commercial organization. All students experienced that these topics were 'less relevant.' Oddly, the topic of social skills was examined by a theoretical test. During his career, Jurriaan realized that these 'soft' skills were treated the same way in most organizations. Especially in organizations with a Human Resources (HR) department, people are seen as resources, not humans. However, Jurriaan is convinced that whenever someone enters a new project, one major success factor is their ability to interact with people. But these topics are underestimated.

When Jurriaan started giving safety training, he noticed that when organizations hired him to provide it, they expected standard training with guaranteed measurable safety improvement. Safety was something of the safety department (supposably where all the wisdom could be found). For him, the question in all the projects should be: *What is necessary and should be done to improve safety in an organization?*

SIMILARITIES IN MANAGING MODERN SOCIETY AND MANAGING SAFETY

In 2003, Jurriaan met Jaap Peters, who wrote the book *De Intensieve Menshouderij* (The Intensive Human Farming), a book on how organizations are organized in a modern society (Peters & Pouw, 2005). Peters & Pouw describe that today's organizations are organized according to Anglo-Saxon principles. This means employees are losing freedom as management determines what the employees ought to be doing. To describe this 'intensive' way of working, Peters uses a metaphor that illustrates the relationship between how intensive agriculture is organized. Intensive agriculture is a method of farming that maximizes yields by using large inputs of capital, labor, and technology. This type of agriculture often involves large-scale farming operations that prioritize efficiency and productivity over sustainability and environmental conservation.

The authors used intensive agriculture as a metaphor for the Anglo-Saxon way of working because they believed that, like intensive agriculture, it focuses on maximizing productivity at the expense of long-term sustainability. In intensive agriculture, farmers use large amounts of chemicals, pesticides, and fertilizers and harvest crops in quick succession, all of which deplete the soil and harm the environment. Similarly, the Anglo-Saxon way of working emphasizes short-term gains, cost-cutting measures, and efficiency at the expense of quality and employee well-being. This can lead to burnout, high turnover rates, and a workplace culture of fear and mistrust. Peters describes that organizations should adopt a more sustainable approach to work that focuses on long-term success and the well-being of employees.

The parallels between intensive agriculture and the Anglo-Saxon way of working are striking: organizations are managed similarly to intensive agriculture.

Peters also wrote several books about Rhinish organizing, such as 'The Rhineland Way.' The Rhineland Way involves all stakeholders and focuses on building and upholding a collective ambition, long-term planning, and collaboration between workers and management. It values sustainability, flexibility (dealing with the unknown), social responsibility, and profitability.

A COLLECTIVE AMBITION TOWARD ORGANIZATIONAL GOALS

The Rhineland approach assumes that people have a collective ambition that needs to be realized. All employees understand why the organization is in operation. For example, Jurriaan is now working with an elderly care organization. Their collective ambition is to facilitate a good life for their clients, and all involved contribute to this collective ambition. The collective ambition is realized through 'working communities.' These working communities consist of the elderly, caregivers, cleaners, family members, doctors, volunteers, and IT support; they all contribute to this collective ambition.

The Rhineland Way of organizing is based on three principles: Connection (invest time to build a community), Trust (trust that people do what is necessary to fulfill the collective ambition), and Craftsmanship (focus on continuous improvement of craftmanship; Figure 3.2).

Jurriaan linked this Rhinish worldview to the world of safety and safety training. He realized that if the future isn't predictable, people need to be prepared for things they didn't expect and must develop resilience. That means people must be trained to think for themselves rather than conducting safety training and following orders from superiors. It is not about receiving a stamped certificate. It requires a

FIGURE 3.2 Rhineland's way of organizing work to achieve the collective ambition.

different way of training people. He personally experienced the effects of the different approaches in training within organizations. He noticed how organizations are organized – Anglo-Saxon or Rhinish – impacts their workforce and lays the foundations of safety behavior on the work floor.

An organization based on Rhinish principles starts with the collective ambition of the organization, defining why the organization exists. They then form a horizontal structure, focusing on where and with whom the work is done (building a working community where everybody joins to reach the collective ambition within and outside the organization). In the horizontal organization, they define who the frontline is and agree on how the back office supports the frontline in reaching the collective ambition.

Employees working for an Anglo-Saxon or Rhinish organization perform the same job and activities but from a different perspective. Jurriaan explains that when the focus is on money in favor of shareholders, the result could be sizable profits for companies and, at the same time, contribute to climate change, pollution, and destruction of the natural environment. The relevance to safety concerns the attitude that comes from the focus on only money as a value. Employees do not feel responsible for the counter effects of their work. They strictly follow rules and procedures.

The effect of a Rhinish way of working is noticeable in how people experience their work. Jurriaan sees that with the Rhineland Way of working, employees tend to have much more self-esteem, take responsibility for their actions, and even rate their jobs as more enjoyable and take less sick leave. When the frontline workers are facilitated in the Rhinish way to perform their jobs, they are triggered to do so at the top of their abilities. They will come up with creative solutions at the time when they need it. It doesn't have to pass by another decision-maker or manager; they will solve the issue themselves. When managers don't have the opportunity to intervene, issues are solved much faster and more efficiently. For example, during the COVID period, hospitals were run by frontline workers while the managers were ordered to stay at home. The resilience of the rest of the frontline workers led to innovative solutions for many unexpected problems that occurred.

THE CONFINED CHICKEN

It's an illusion to think that an organization can easily switch from the Anglo-Saxon way of organizing work to the Rhineland Way. Some managers interpret the latter approach as sending the message to the frontline workers, saying, '*From now on, you're a self-steering team*,' and then observe them, realizing after a while that the team isn't performing well. Jurriaan compares it, ironically, to a chicken: when a chicken has spent most of its life in a confined space and is suddenly released into an open field, the chicken will be confused, as all of a sudden, their food is not at hand, the space around them is scaring them, and most chickens don't survive this surge of freedom.

The Anglo-Saxon model assumes that rules and procedures are written for a reason and need to be performed precisely how they are written. The Rhineland Way of thinking assumes reasonableness and fairness while doing the job. Jurriaan concludes by saying that to achieve resilience and a safe way of working, you need

The Cynefin model is a sense-making framework used for decision-making by identifying the nature of a problem or situation and then applying appropriate management and decision-making techniques. It was developed for use in knowledge management and organizational strategy.

The framework divides problems into five categories: Simple, Complicated, Complex, Chaotic, and Disorder. The categories indicate the degree of uncertainty involved in the problem and the appropriate management approach:

- Simple problems are straightforward and are best addressed using a sense and respond approach.
- Complicated problems require greater analysis and expertise and involve multiple factors that must be considered before identifying a solution.
- Complex problems involve many interconnected factors and involve a level of unpredictability. This requires an approach that is iterative and adaptive.
- Chaotic problems emerge due to unforeseen circumstances and are best addressed through action and experimentation.
- Disorder category represents the state of not knowing which of the four domains the situation is.

(Kurtz & Snowden, 2003)

people that work according to the Rhinish worldview: 'Spreadsheets don't make the world safer, Rhinish people do.', according to Jurriaan.

COMPLEX VERSUS COMPLICATED

The difference between the two approaches is that the Anglo-Saxon model assumes a complicated system, and the Rhineland Way of thinking is a complex system. A complicated system can't function with simple and creative solutions; there are too many factors to be considered, and many people need to be involved. For those who recognize something familiar, the Cynefin framework describes problems as either being complex or complicated, as well as three other categories, namely simple, chaotic, and disorder.

MANAGING SAFETY

A safety management system (SMS) is a systematic approach to managing safety, as has been introduced in Chapter 2. The concept of SMS was first introduced in the aviation industry in the late 1980s by the International Civil Aviation Organization (ICAO) as a way to improve safety in the aviation industry by focusing on proactive risk management rather than reactive responses to accidents.

The Civil Air Navigation Services Organization (CANSO, the global voice of the air traffic management [ATM] industry) significantly emphasizes safety culture within an SMS. Rather than identifying safety culture as a separate element, they place it as an overarching or integral part.

According to the Occupational Safety and Health Administration (OSHA, 2016), an effective SMS can help identify and control hazards, reduce accidents and injuries, improve morale and productivity, and reduce costs associated with accidents or incidents. The International Labour Organization (ILO, 2001) emphasizes that investing in safety is not only a legal obligation but also a moral responsibility of employers to protect their employees from harm. Investing time in establishing an SMS contributes to protecting employees, complying with regulations, reducing costs associated with accidents or incidents, improving efficiency, enhancing organizational reputation, and fulfilling moral obligations.

Critics of SMS argue that an SMS may not always be effective in preventing accidents or improving safety. Some suggest that SMS can become bureaucratic and overly focused on compliance rather than proactive hazard identification and risk management (Dekker & Nyce, 2018). There are concerns about the potential for SMS to become 'tick-box' exercises that prioritize paperwork over practical safety measures (Linnenluecke & Griffith, 2010). Additionally, some point out that SMS may not address underlying organizational factors such as culture or leadership that can contribute to accidents (Reason, 1997).

THE S-WORD

As many introductions about safety culture will describe, the term 'safety culture' was introduced in the 1980s by the International Nuclear Safety Advisory Group (INSAG, 1996) following the Chernobyl disaster. INSAG recognized that technical failures alone were not enough to explain the disaster and that underlying cultural and organizational factors were at play. They defined safety culture as 'the product of individual and group values, attitudes, perceptions, competencies, and patterns of behavior that determine the commitment to, and the style and proficiency of, an organization's health and safety management.'

The concept of safety culture has since been applied – and it is up to debate what that means – to various industries beyond nuclear power plants, including aviation, healthcare, and manufacturing. A positive safety culture emphasizes the importance of creating a workplace where safety is prioritized at all levels of the organization and where employees feel empowered to report safety incidents or concerns without fear of retaliation.

SAFETY CULTURE

Despite abundant literature on the topic, researchers have not yet reached a consensus on the precise definition, assessment, and management of safety culture. Even though the concept has matured since its inception, the debate continues to this day and includes the confusion about whether it should be called safety culture or safety climate. These two concepts have been used interchangeably in the literature, leading to academic arguments about which to use for research on organizational safety.

Reason (1998) has suggested that safety culture consists of five elements:

- An informed culture: Those who manage and operate the system have current knowledge about the human, technical, organizational, and environmental factors that determine the safety of the system as a whole.
- A reporting culture: Managers and operational personnel freely share critical safety information without the threat of punitive action.
- A learning culture: An organization has the willingness and the competence to draw the right conclusions from its safety information system and the will to implement major reforms.
- A just culture: An atmosphere of trust in which people are encouraged to provide essential safety-related information but in which they are also clear about where the line must be drawn between acceptable and unacceptable behavior.
- A flexible culture: A culture in which an organization can reconfigure itself in the face of operations or certain kinds of danger – often shifting from the conventional hierarchical mode to a flatter mode.

While there are several categories of safety culture in the literature, the most commonly (Hudson, 1999) used ones are:

- Pathological: We don't care about safety as long as nobody finds out.
- Reactive: We care much about safety; we do a lot after each accident.
- Calculative: We have safety systems in place to manage hazards.
- Proactive: We deal with the safety problems that we still encounter.
- Generative: We brainstorm together to be a step ahead of safety issues.

PROBLEMS WITH SAFETY CULTURE

Despite the attributed importance of safety culture in promoting safe work practices, there are still several problems that organizations might face regarding this topic. Throughout the years, academia and industry have voiced criticism of its definition, application, and overall usefulness in understanding and improving safety.

One issue is the lack of clear definitions and metrics for safety culture, which makes it difficult to measure and improve. Guldenmund (2000) found that many organizations struggle with defining safety culture and interpreting its implications for their operations. Another problem is the tendency for safety culture initiatives to be driven by compliance rather than a genuine commitment to safety.

CLIMBING THE 'LADDER'

It is no secret that Carsten Busch – the 'Indiana Jones of safety' – is critical of the term 'safety culture.' Especially the introduction of the safety culture ladder drove him to start questioning what was happening surrounding the topic. In essence, there is a reasonable model to talk about and discuss how safety could be improved within the organization, but why did it have to become an ISO-like (International Organization for Standardization) standard? Also, why are these standards now mandatory, as organizations have established contractual requirements for contractors?

'Organizations have abundantly started to use the Safety Culture Ladder in places where it is not applicable,' Carsten says.

THE FIRST RULE OF SAFETY CULTURE

Carsten wonders what added value if the word 'culture' is in this context. One could also change the word 'culture' to 'management' and that would still mean the same thing. In fact, one could opt for leaving out both 'culture' and 'management' – which would lead to an organization working on their *safety* – and that would *still* mean the same thing, if not give the sentence more meaning by providing a more specific context.

> The focus and efforts shouldn't be towards safety culture. The focus should be on the bigger picture – safety – and should include specific behavioral change intentions.
>
> *– Carsten Busch*

The overuse and misuse of 'safety culture' annoyed Carsten in such a way that he decided to write a book about the topic: *The First Rule of Safety Culture* (Busch, 2021) – referring to the movie *Fight Club*, in which the first rule of fight club is to 'not speak about fight club.'

Carsten believes that culture as a concept is not the problem – he underlines that group cultures within organizations do exist. Group cultures indeed hold aspects of safety; therefore, for ease of use, we might as well call it safety culture.

However, this doesn't mean the focus and efforts should be toward safety culture. The focus should be on the bigger picture – safety – and should include specific behavioral change intentions. For example, Carsten mentions that he knows of managers – say HR managers – who've stated that there will be no culture change projects or programs within their organization. These organizations reject culture-focus in favor of focusing on how they do the work, and they believe a change in culture will follow, as culture is the result of how people work and interact, according to Carsten.

> Measuring safety culture is delivering baloney in an organized manner, similar to ISO 9001.
>
> *– Carsten Busch*

Accordingly, measuring safety culture is 'delivering baloney in an organized manner, similar to ISO 9001,' Carsten says. He compares safety culture assessments to personality assessments during job application processes: The only function safety culture assessments have, according to Carsten, is a reason to discuss all things safety within the organization afterwards, 'but not to benchmark, climb any ladder, or consider the results as set in stone.'

SAFETY CULTURE: MEASUREMENT AND LIMITATIONS

Jaap van den Berg (Safety Culture Manager in Aviation) mentioned that not every airline uses the safety culture ladder as a measurement tool for safety culture. He is, however, aware that Southwest Airlines, for example, has used it and that they scored the highest step, which is the generative safety culture. Probably not coincidentally, they also score very high on passenger satisfaction.

The employees at Southwest Airlines experience a high degree of autonomy in their jobs. For example, one senior cabin crew member decided to skip the traditional safety briefing and bring a twist to it. He asked the passengers to start clapping to a certain beat. He started rapping out the safety briefing. All the passengers in that airplane were clapping along, and passengers were engaged in this safety briefing. Word got out about this crew member, and top management decided it was time to take action. They invited the rapping cabin crew member to their annual sharehold-ers' meeting and rap the annual report in front of the shareholders rather than the usual program. Management saw their unique disposition through their employees and made it work for them. This reflects their company culture.

In contrast, a similar thing happened in Germany where a German senior cabin crew member deviated (perhaps slightly?) from the usual safety briefing, and this crew member got fired because it was not compliant. This example represents the complexity and variety of various cultures at play: organizational, professional, and national culture.

Although measuring safety culture has potential benefits, it's essential to also con-sider potential drawbacks. The benefits of measuring safety culture include providing a clear and structured approach for improving safety culture, aiding organizations in identifying areas for improvement and tracking progress over time, encouraging involvement and engagement from all levels of the organization, and possibly leading to improved safety outcomes and reduced incidents. However, it's difficult to mea-sure, and research on this topic is inconclusive.

There are several limitations to measuring safety culture. Firstly, it may be per-ceived as too rigid or inflexible, which could lead to employee resistance or lack of engagement. Additionally, the assessment process can be quite lengthy and require significant resources. Another concern is that some organizations may prioritize climbing the ladder over addressing underlying safety concerns. Finally, some experts have raised doubts about the ladder's criteria and whether they accurately reflect the concept of a safety culture.

Organizations may become overly focused on climbing the safety culture ladder rather than addressing actual safety concerns. This ladder-climbing mentality allows organizations to demonstrate their commitment to safety without making the neces-sary changes to ensure it. Essentially, it is a way of 'doing safety' without actually 'doing safety.' This approach is all too common in today's corporate world, where optics and appearances often take precedence over real action.

The approach of measuring and 'climbing' can lead to a false sense of safety, where organizations may believe they are doing enough to ensure safety when they are not. To truly improve safety within organizations, they must move away from the ladder-climbing mentality and focus on making the necessary changes to ensure the safety of all employees.

PLANNING (FOR SAFETY CULTURE) PREVENTS POOR PERFORMANCE

The topic of safety culture has been discussed in both academia and industry for a while now. However, there's still little agreement on whether safety culture reflects how an SMS operates or the effects of such a system on safety performance. In addition, organizations don't *consistently* assess safety culture.

Researchers (some of my esteemed former colleagues and myself) at the Aviation Academy of Amsterdam University of Applied Sciences proposed a framework for developing a safety culture based on academic literature and industry standards, following Reason's (1998) description of safety culture (Karanikas et al., 2015). Kaspers et al. (2016a, b, c). The framework, named the Aviation Academy Safety Culture Prerequisites (AVAC-SCP), subdivides safety culture into 37 prerequisites across six areas:

1. General prerequisites
2. Just culture
3. Flexible culture
4. Reporting culture
5. Informative culture
6. Learning culture

The framework can help demonstrate the gap between current conditions in an organization and ideal conditions where all conditions are fully implemented. What's different about this approach is that it uses existing academic theory but not to measure objectively against a fixed measurement tape; it measures whether organizations have described safety policy, whether they have implemented described policies, and whether employees see these efforts in their day-to-day job.

The bigger the gap between these three conditions, the more organizations need to focus on closing these gaps. If the organization has a safety policy in place, but management hasn't implemented it or has implemented it, and employees don't see how, it becomes clear what activities are focus points for improvement. Ideally, there should be a match with how work is described, how work is done, and how work is imagined.

WORK-AS-DONE AND WORK-AS-IMAGINED

Measuring gaps between an envisaged situation and reality has been named Work-as-Done and Work-as-Imagined. Hollnagel (2014) describes Work-as-Imagined as a theoretical view of the formal task environment that assumes ideal working conditions, ignoring the reality that tasks must be adapted to changing conditions in the work environment and the world. It describes what should happen under normal circumstances, while Work-as-Done describes what actually occurs and how work progresses over time in complex contexts.

Current safety culture measurement tools have some notable limitations (Piric et al., 2017): they are mostly not grounded on sound theoretical frameworks; they tend to measure perception rather than factual information; they provide a result based on a scale (usually from 1 to 5), suggesting that the perceived safety culture can be graded and suggest there is a 'finite' grade to be achieved after which the organizations can stop developing their safety culture.

The AVAC-SCP measures something else:

a. Does your organization have the documentation in place in which planning for safety culture is visible?
b. Has the organization implemented what it said it would do under a)?
c. Do the employees see these above-mentioned efforts at their workplace?

Figure 3.3 shows what the results might look like for various organizational departments and where gaps between policy, implementation, and perception can be observed. In Chapter 9, we will dive deeper into the use of the tool.

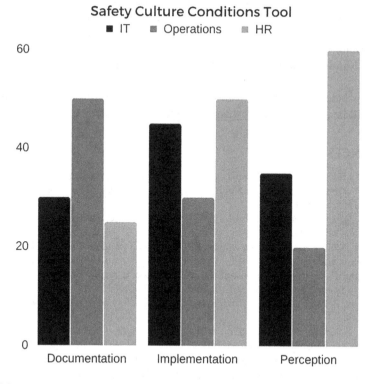

FIGURE 3.3   Example result of the AVAC-SCP tool on policy, implementation, and perception for various organizational departments.

CONCLUSION

Although significant progress has been made in the last two centuries, it appears that organizations are still in the process of understanding how to best manage work, employees, and safety in general. Although existing theories, models, and approaches have certainly progressed our thinking about how safety should be managed, the field is imperfect and subject to refinement in application and measurement. In managing safety, an organization's safety culture is often considered, but it remains a controversial topic with benefits and limitations. Some concluding take-away thoughts of this chapter are:

1. **Research** the current organizational management approach to work and find out whether this approach fits the organizational goals, mission and vision, and the employees working toward those.
2. **Keep informed but don't selectively disregard:** Theory on safety is in constant motion and new concepts, applications, and approaches arise. Safety theorists disagree often, but that doesn't mean the field is black and white: keep informed about what's going on and keep an open mind as to what works for you and your organization.
3. **Don't be scared to measure:** There are many tools out there that can be used for various goals.
4. **Do know the limitations and side effects** of measurements.

As we conclude this chapter and move onward, the topic of measurement and performance will be discussed more extensively in the next chapter.

REFERENCES

Busch, C. (2021). *The first rule of safety culture: A counter-C-word manifesto.* Mind The Risk.

Dekker, S., & Nyce, J. M. (2018). *Safety differently: Human factors for a new era* (2nd ed.). Boca Raton, FL: CRC Press.

Federal Aviation Administration (FAA). (2020). Safety Management System Implementation Guide. Order 8000.369B.

Flanders, J. (2003). *The Victorian house: Daily life from childbirth to deathbed* (p. 106). London: HarperCollins.

Guldenmund, F. W. (2000). The nature of safety culture: a review of theory and research. *Safety Science*, 34(1–3). doi: 10.1016/S0925-7535(00)00014-X.

Guldenmund, F. W. (2010). Understanding and exploring safety culture. *Safety Science*, 49(2), 55–64.

Hollnagel, E. (2014). The construction of Safety-II. In *Safety-I and safety-II: The past and future of safety management* (pp. 120–123). Aldershot: Ashgate Publishing.

Hudson, P. (1999). Safety culture – theory and practice. Presented at the workshop on "The human factor in system reliability – is human performance predictable?" Siena, Italy, 1–2 December 1999.

International Atomic Energy Agency Safety Report (INSAG). (1996). INSAG-7: The Chernobyl Accident, Updating INSAG-1. Retrieved from: http://www-pub.iaea.org/MTCD/publications/PDF/Pub913e_web.pdf

International Civil Aviation Organization (ICAO). (2006). Safety Management Manual (SMM). Doc 9859 AN/474.

International Labour Organization (ILO). (2001). Introduction to Occupational Safety and Health: An ILO Perspective. https://www.ilo.org/wcmsp5/groups/public/---ed_protect/---protrav/---safework/documents/instructionalmaterial/wcms_107870.pdf

Karanikas, N. (2015). Correlation of changes in the employment costs and average task load with rates of accidents attributed to human error. *Aviation Psychology and Applied Human Factors*, 5(2), 104–113. doi: 10.1027/2192-0923/a000083.

Kaspers, S., Karanikas, N., Roelen, A. L. C., Piric, S., & de Boer, R. J. (2016a). Review of Existing Aviation Safety Metrics, RAAK PRO Project: Measuring Safety in Aviation, Project S10931. Aviation Academy, Amsterdam University of Applied Sciences, the Netherlands.

Kaspers, S., Karanikas, N., Roelen, A. L. C., Piric, S., van Aalst, R., & de Boer, R. J. (2016b). Results from Surveys about Existing Aviation Safety Metrics, RAAK PRO Project: Measuring Safety in Aviation, Project S10931. Aviation Academy, Amsterdam University of Applied Sciences, the Netherlands.

Kaspers, S., Karanikas, N., Roelen, A. L. C., Piric, S., van Aalst, R., & de Boer, R. J. (2016c). Exploring the diversity in safety measurement practices: Empirical results from aviation. Proceedings of the 1st international cross-industry safety conference. *Journal of Safety Studies*, 2(2), pp. 18–29. Amsterdam, 3–4 November 2016. doi: 10.5296/jss. v2i2.10437.

Kurtz, C. F., & Snowden, D. J. (2003). The new dynamics of strategy: Sense-making in a complex and complicated world. *IBM Systems Journal*, 42(3), 462–483.

Linnenluecke, M. K., & Griffiths, A. (2010). Beyond compliance: The limits of the safety management system approach. *Journal of Air Transport Management*, 16(4), 181–186.

Occupational Safety and Health Administration (OSHA). (2016). Recommended Practices for Safety and Health Programs. Retrieved from: https://www.osha.gov/shpguidelines/docs/OSHA_SHP_Recommended_Practices.pdf

OECD. (2004). Observations on the Anglo-Saxon Model of Corporate Governance. The OECD Observer. Retrieved from: https://www.oecd.org/corporate/ca/corporategovernanceprinciples/31557724.pdf

Peters, J., & Pouw, J. (2005). *Intensieve menshouderij: hoe kwaliteit oplost in rationaliteit.* Nederlands: Scriptum.

Piric, S., Karanikas, N., de Boer, R., Roelen, A., Kaspers, S., & van Aalst, R. (2018). How much do organizations plan for a positive safety culture? Introducing the Aviation Academy Safety Culture Prerequisites (AVAC-SCP) Tool. Proceedings of the international cross-industry safety conference, 2–3 November 2017, 1(1), pp. 118–129. Amsterdam University of Applied Sciences, AUP Advances. doi: 10.5117/ADV2018.1.008.PIRI.

Reason, J. (1997). *Managing the risks of organizational accidents.* Aldershot: Ashgate Publishing.

Reason, J. (1998) Achieving a safe culture: Theory and practice. *Work and Stress*, 12, 293–306.

Snowden, D. J., & Boon, M. (2006). A leader's framework for decision making. *Harvard Business Review*, 85(11), 68–76.

4 Counting Definitely Counts, But Not Always

> Not everything that can be counted counts and not everything that counts can be counted.
>
> *– Albert Einstein*

THE DOUBLE-EDGED SWORD OF SAFETY PERFORMANCE INDICATORS

As seen in the previous chapter, safety assurance is one of the pillars of a Safety Management System (SMS). It describes that safety indicators should be monitored, and safety performance should be assessed over a given period.

The only problem is, how do you measure safety? When I ask my students to answer this question, they usually come up with 'the number of air crashes,' 'the number of terrorist attacks,' or 'the number of *some other catastrophic events*.' Yes, you can measure these events, but if you haven't had one in 50 years, are you confident that the operation is safe?

In 2008, Transocean – which owned the Deepwater Horizon drilling rig – received the SAFE award from the US Minerals Management Service for its 'outstanding drilling operations' and 'perfect performance period.' In 2010, the Deepwater Horizon drilling rig exploded and sank in the Gulf of Mexico. The accident resulted in the death of 11 workers and caused one of the worst environmental disasters in US history, with millions of barrels of oil released into the sea. It shows that even when you measure safety and apparently do well in managing safety, it doesn't guarantee safety for the future.

Although safety is difficult to measure, one can try endlessly. A wide range of indicators can be used, and since you probably already work in the safety domain, you are already familiar with a few of them. But how do companies know whether their efforts are effective? What is the added value of all the data that they are gathering? And what if your organization is too small to measure large data sets?

THERE IS NO SUCH THING AS A RELIABLE SAFETY PERFORMANCE INDICATOR

Global Health, Safety, Environment & Quality (HSEQ) Director Kick Sterkman works for an Upstream Oil & Gas Company. He was responsible for safety as an offshore manager for installations in various regions, including the North Sea, The Netherlands, Germany, and Norway. However, he found himself investigating the same incidents repeatedly, such as the recurring occurrence of someone losing a toe due to the use of a high-pressure washer. This sparked his curiosity and motivated him to seek answers to these questions. To gain insight, Kick began to read

DOI: 10.1201/9781003383109-4

and connect with safety professionals in the offshore industry and even obtained a Master's degree in Management, focusing on Psychology. Though the psychology modules provided some understanding of human motivation, they did not give the answers Kick was searching for. To complement his knowledge, he also obtained a Master's degree in Human Factors & Systems Safety. However, he realized that the answer to his questions may not exist, as it is a complex mixture of various factors. It is also challenging to identify which factors correlate with the consequences.

Within the oil and gas industry, there was a major project that looked into leading key performance indicators (KPIs), and Kick said that the main conclusion of that work is that none of the leading KPIs were found to consistently correlate with an effect. The question is, then, what can you do next?

OUTSOURCE THE COMPLEXITY OF SAFETY METRICS
TO A PARTY THAT UNDERSTANDS THEM

Despite the disappointing results in the Leading KPI project, the team did not want to give up and found new inspiration in the film *Moneyball*. Following these insights, they concluded that the variations in KPIs, such as the total recordable injury rate (TRIR), are not so, if at all, statistically significant that they can be correlated to improvement or deterioration of the SMS. From a safety management perspective, this is deemed important to avoid the endless series of narratives that serve to explain the variations in TRIR to the organization and that ultimately undermine the credibility of safety messages.

Starting from the assumption that there must be an underlying probability of a recordable incident occurring because of the organization's operational footprint and activities, the search for this parameter started. This search required a highly technical data analysis which was outsourced to an external specialist company. From over ten years of detailed incident statistics, an indicator was distilled that represented the underlying probability of a recordable incident in the organization. The company now uses this metric in a dashboard that displays the 12-month rolling TRIR and visualizes where it resides within the calculated confidence bands. This, then, enables the understanding if the variations meet the expected randomness or are more likely to represent a change in the underlying probability.

In essence, the underlying probability is both a lagging KPI, as it is based on historical data, as well as a leading KPI. A leading KPI in the sense that it allows to make predictions about the future TRIR values with a certain statistic accuracy. Unfortunately, statistics cannot provide insight into when these incidents will happen; if anything, this work shows that the occurrence of these types of incidents is random within an established organization.

LEADING AND LAGGING INDICATORS

Safety performance indicators are classified as leading or lagging indicators. Lagging indicators measure safety events that have already occurred, while leading indicators can be used to prioritize safety management activities and determine actions for safety improvement (Roelen & Papanikou, 2020).

Regarding indicators, the next question is what factors impact the likelihood of incidents occurring. Determining what truly influences these chances and what doesn't can be challenging. Kick's perspective is that every bit of knowledge and theory related to safety can contribute to a more complete understanding of the issue. He's noticed that academic papers tend to focus on new safety insights and disregard past knowledge, but Kick believes that building upon existing knowledge incrementally is more effective. In his opinion, basic safety measures such as permits, procedures, and technical safety systems are essential, but new safety thinking should also be implemented on top of these measures. Combining old and new knowledge is vital to achieving a comprehensive safety approach.

Old knowledge and the new views should work in conjunction.

– Kick Sterkman

NEW VIEW OF SAFETY

Hollnagel (2014) argues that to effectively learn and improve safety in organizations, we must move beyond simply reacting to adverse events and instead focus on understanding the processes and decision-making that led to those outcomes. This new view of safety, also called Safety-II, focuses on a system's ability to adapt and succeed under varying conditions. It recognizes that everyday performance variability can help a system respond to different circumstances and be resilient. In this view, humans are recognized as a valuable resource for a system's flexibility and adaptability. Instead of trying to understand why things occasionally go wrong, investigations under a Safety-II approach seek to understand how things usually go well. The traditional approach of categorizing events as 'good' or 'bad' and searching for a single 'root cause' or blaming 'human error' is too simplistic in modern, complex systems where people often have incomplete information and competing goals. Instead, Hollnagel proposes that we shift our focus to learning about the work process, regardless of the outcome (Figure 4.1).

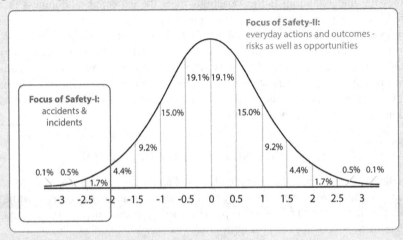

FIGURE 4.1   Two types of focus of safety.

As an organization, it takes courage to declutter data and eliminate anything that doesn't contribute to gaining valuable insights. The Upstream Oil & Gas Company collaborates with a shared incident management database within the sector that identifies common root causes of incidents. These root causes may include inadequate supervision, failure to follow procedures, or insufficient training.

The question is, why doesn't the sector address these root causes and eliminate these safety issues? The problem lies in what Kick calls the maze effect: once you're in a maze, you can't look around the corner to see what happens next. When you start at the end of the maze and work backward, there is only one way, and you can connect the result with the root cause – the beginning of the maze.

Classical hindsight bias: When you start at the beginning of the maze – the root cause – there are a million possibilities, but only one leads to the result, and then the question is why. One method that allows the organization to create opportunities to discuss the moments between the employee's actions and the incident is called restorative justice.

RESTORATIVE JUST CULTURE

The Restorative Just Culture approach aims to repair trust and relationships that have been damaged after an incident. It involves allowing all stakeholders to discuss how they have been affected and working together to decide the best ways to repair the harm caused. Stakeholders are described as first and second victims, organizations, community, and others that need to be specified. The approach aims to detail the needs of all stakeholders to enable accepting responsibility, emotional healing, the reintegration of the people involved, and to address the causes of the incident (safetydifferently. com, 2023).

The Upstream Oil & Gas Company Energy has started to use the benefits of artificial intelligence to derive meaningful predictions. The system can make predictions for future operations by looking at the circumstances of previous events or incidents. When that circumstance is similar, it becomes highlighted, and operational personnel is reminded of the events surrounding the previous incidents. For example, suppose they know that a specific type of work will be performed at a particular time of the day by a specific unit. In that case, the system provides information on incident probability. This probability is then compared to the normal or average incident probability and forms the basis of a briefing for a shift or a project to highlight certain risks. By using data in this way – to provide meaningful insights into safety – operational personnel can be facilitated in their work and safety by drawing extra attention to the circumstance during their work and making them aware (again) of the risks.

ACCOUNTABILITY COMES WITH AUTONOMY

Real organizational learning comes from identifying what someone has done and, more importantly, understanding *why* this person did it in that particular way. In their company values, the Upstream Oil & Gas Company has included that all employees are accountable for their actions, which means that they won't be punished for their actions, but it is expected of them to explain what they have done and why if they were part of a sequence of events. What is inherently tied to this value is that employees should have the autonomy to decide how they can do their work best. There is no point in giving employees strict work instructions and making them accountable if it doesn't work out. Operational personnel need to experience freedom in decision-making if they are also held accountable for their actions. It will allow them to become the owner of the process. The operational line is responsible for safety, and according to Kick, this is imperative to the positive working of safety culture within the organization. Making operational personnel accountable has enabled safety to become a topic of importance for them. This is the first reason he thinks the Upstream Oil & Gas Company is successful in their HSE.

> There is no point in giving employees strict work instructions and then making them accountable if it doesn't work out – process follows people.
>
> – *Kick Sterkman*

Safety Performance Indicators: A Useful Gauge If You Know What You're Doing

The second factor for this success is somewhat controversial, and there have been discussions among safety professionals of the (dis)proven effect of it: the incentivization of safety.

At the Upstream Oil & Gas Company, employees are coupled to personal performance and a safety scorecard. These scorecards represent a country or a group of countries and can make up to 25% of a bonus. That means the bonus can be paid when a group or country performs well safety-wise. When performance is measured at such a high level of groups or countries, do employees still see their personal contribution to safety? Kick says they do because employees work directly with their colleagues on the installations in their country and are responsible themselves during their work as well.

But will employees sweep inevitable mistakes or incidents under the rug? That's not possible, according to Kick. In their business intelligence model, they check for spillovers or a higher variation in other indicators. For example, they've (re-)introduced certain recordables, such as the provision of first aid.

The second measure they've taken to avoid people not reporting is the introduction of reporting frequency: how many reports are filed per million man-hours, which are benchmarked between the nations. They put a lot of communication effort into ensuring the reporting rate is as high as possible.

What's remarkable about these reports is that they receive safety observations and reports on positive situations. In other words, in line with Safety-II thinking, the organization can see what goes well.

The third measure they've taken is that the Upstream Oil & Gas Company Energy has outsourced their medical aid to an external company. They are the ones that judge each incident on whether the employee requires medical treatment or if first aid would suffice. The organization does not influence the judgment of the care provider.

Kick knows this topic is incredibly controversial, but he believes incentivization is the secret behind their safety success.

GAMING THE SYSTEM: TIPP-EX INCIDENTS

At the opposite end of the incentivizing safety spectrum is Bart Vanraes' research into Tipp-ex incidents. Bart is an independent occupational prevention advisor. Tipp-ex accidents are occupational accidents that are not fully reported; organizations decide to hide the accident under a proverbial layer of Tipp-ex. In other words, it is not only the under/non-reporting but also the downgrade of the occurrence. For example, a serious fracture that is classified as the provision of first aid. There is a paper trail, although the report is filled in 'creatively' and the occurrence might not appear on any dashboard.

INSULT OVER INJURY

With Tipp-ex incidents, the incidents are concealed. Usually, these are incidents with significant consequences, such as severe injuries causing people to lose time at work. This is a much bigger problem than most of us realize; according to Bart, it is a global and underestimated problem. In a blog post (Vanraes, 2020), Bart mentions that the Dutch labor inspection estimated that 50% of occupational accidents (lost time incident/accidents) are not reported (Nederlandse Arbeidsinspectie, 2021).

UNMASKING THE TRUE PICTURE OF SAFETY

Safety dashboards might give the illusion that we are entirely in control of our safety performance. The reality is that when we see a specific trend go down, we relax and allow ourselves to breathe, but when a trend goes up, we suddenly sit at the edges of our seats, even though, in both instances, we probably don't know why the trend goes up or down. The natural tendency is to go out and seek more safety-related data, despite the possibly limited contribution to meaningful insight. It's easy to lose sight of the details of data because, from data, it is simply unknown why certain events happen.

It can be challenging to find meaningful indicators for safety in certain situations. One example from healthcare of how safety data didn't contribute to patient safety is explained by Miriam Kroeze (Senior Medical Advisor at MediRisk). At one point, hospitals wanted to tackle the number of postoperative wound infections with patients after surgery. The way to contract this infection is by the air quality in the operating room, which is influenced by – among other factors – the temperature and the number of people in the room. In the operating room, many people are present during surgery, not just surgeons and nurses but also trainees. During surgeries, it is sometimes necessary for them to leave the room for various reasons, such as shift change, to attend to emergencies, to get medical equipment, and coffee breaks. These movements within the operating room must be minimized to ensure the quality of the operating room.

What would be a good indicator of air quality within the operating room? The hospitals came up with the number of times the operating room door was opened during surgery as an indicator. The thought was that if the number of times the door was opened increased, the air quality would drop and, therefore, should be as low as possible. Hospitals decided to steer on this particular indicator, and they attached a target of a maximum of three times during surgery, a set norm per type of operation. The door can remain closed for simple and quick operations, while the maximum standard may be slightly higher for complex or longer operations. Counting start at incision and stopped at wound closure. Soon various gadgets and monitors were installed by companies to show the healthcare staff whether the amount of time was 'in the green' or if it had reached its limit and 'red.'

> The consequence was that healthcare staff started to perform all sorts of workarounds that didn't facilitate them in their work during surgery just to be able to minimize the performance indicator.
>
> *– Miriam Kroeze*

The indicator didn't serve its purpose and even became dangerous. It wasn't the number of times the door was opened that was the real issue; it was the number of movements within the room that was the issue. Sometimes, staff needed to leave the room and come back to be able to facilitate the surgery, for example, if they needed specific instruments that weren't at hand. The consequence was that healthcare staff started to perform all sorts of workarounds that didn't facilitate them during surgery just to minimize the number of times the door was opened.

This example shows that the world is complex in a way we can't always anticipate, properties emerge we couldn't predict, and steering based on flawed indicators or faulty mitigating measures can cause dangerous workarounds. A complex world cannot be managed by linear solutions. Ultimately, MediRisk introduced the functional resonance analysis model (FRAM) to hospitals to make the complexity within the healthcare processes visible and discussable with the care providers involved. FRAM models the work-as-done. As such, it shows the variations and adjustments people make in daily practice. This is about gaining insights to better support healthcare providers in the work they do and not about correcting behavior.

DEFINITION OF FRAM

The FRAM is a model that analyzes complex socio-technical systems. It provides a framework for understanding how accidents happen in complex systems by analyzing the interactions between the different components within the system.

FRAM is based on the idea that accidents are not caused by individual errors or failures but instead arise from the complex interactions between different elements within a system. The model focuses on the functions of different parts within the system rather than their structure or design.

FRAM consists of a set of basic rules and principles describing how elements within a system interact. It also provides a visual representation of the system, which can help to identify potential sources of error or failure (Figure 4.2) (Hollnagel, 2012).

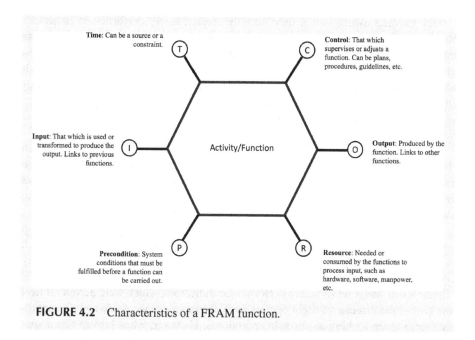

Time: Can be a source or a constraint.

Control: That which supervises or adjusts a function. Can be plans, procedures, guidelines, etc.

Input: That which is used or transformed to produce the output. Links to previous functions.

Activity/Function

Output: Produced by the function. Links to other functions.

Precondition: System conditions that must be fulfilled before a function can be carried out.

Resource: Needed or consumed by the functions to process input, such as hardware, software, manpower, etc.

FIGURE 4.2 Characteristics of a FRAM function.

SHIFTING THE FOCUS: FROM SAFETY METRICS TO PROACTIVE SAFETY

'If the incident reports that a safety department is getting are not resulting in meaningful action,' Nuno Aghdassi, an Air Safety Investigator in Portugal and former Head of Flight Safety at a leading private jet operator, says. 'Then the first question is, are you doing your job properly as the safety department? In other words, are you finding the right things to begin with? Are you looking in the right place? Sometimes people might look at the superficial level and not go down to the root of what is causing the problems.'

'Secondly,' he continues.

'If you're getting these reports and they're not leading to any safety actions, are they the things people should be reporting? It could be that there might be a problem in the reporting culture where people are just reporting problems that happen with other people rather than your organization (such as bird strikes).'

At the private jet operator, most operated aircraft weigh below 27 tons. According to regulations, there was no need to conduct flight data monitoring (FDM), which monitors parameters such as altitude, speed, heading, aircraft attitude, and other in-flight characteristics during the flight. This data provides an organization with information on how the flight crew executed the flight through reconstruction. It's a sure method of improving operational insight, how incidents might develop, identifying potential operational and safety risks, and improving pilot training, among many other benefits.

The private jet operator wanted to implement FDM in all their aircraft. Very soon, they saw the benefits it brought regarding safety, especially in an operation like a business jet operator where things are not routine like at an airline. For example, they could suddenly practice those challenging airports in the simulator they would fly to the next season. In an ever-changing operation such as theirs, they could equally apply the data to other non-standard operations in their format.

One exciting aspect is that they initially created safety performance indicators based on things thrown up from FDM. This was very reactive in a way. To address this, the organization created a series of safety performance indicators about loss of control in flight, for example, related to the state of energy of the aircraft or high bank angles. Concerning runway excursions, they looked at, for example, aircraft lifting off in the last portion of the runway. The result was a set of safety performance indicators, which were proactive in that they were inspired by the key operational issues but did not necessarily mean any present danger.

The organization started to see FDM as something critical to decision-making in the organization. They recognized that the risk owners are not those who work in the safety department; they are a service organization. The risk owners are the chief pilots or directors of operations that are maintenance, and they should be feeding them with everything they can to help them discharge their responsibility in terms of safety.

Chief pilots were asked what their main safety concerns were. They came up with a list of concerns. The organization then created a specific dashboard for each of those, a set of safety performance indicators tailored to that department's safety concerns and interests.

The result was a set of safety performance indicators, which went across all fleet and served the director of flight operations. For example, one can do so in an aggregate form when looking at unstable approaches. However, when broken down specifically to each fleet or type of aircraft, they have specific aspects contributing to safety performance indicators. They were looking at the time it takes to brake on the ground. They customized a set of safety performance indicators specifically for aircraft without thrust reversers.

The key is to help decision-makers, risk owners, and people who deliver safety on the frontline through their decisions and feed them with the necessary safety information.

EXPECT RESISTANCE

When FDM was announced to be implemented, it was received with many reservations by pilots. They thought Big Brother would spy on them during their day-to-day work. After a while, Nuno and his team started to see two major changes: one of them was operational changes in operational behavior. For example, when they detected unstable approaches, with the use of FDM, they were able to raise crew members' awareness, showing them their data. At the private jet operator, pilots would receive snapshots of their records in terms of unstable approaches or long landings every month. This was instrumental in people regulating their behavior; pilots were going to try and improve.

Investigators saw this improvement quite clearly in unstable approaches in training programs. They could use their FDM for training purposes and use scenarios that could then be trained in the simulator.

These efforts have brought about a positive effect. By gaining pilots' trust in FDM, there would be times when pilots would ask to see their own data. They would say, 'Look, this happened on so-and-so flight; can you please show me the data when it arrives?'. Suddenly, pilots showed an active interest, or before they would go into the simulator, they would come and sit down with the FDM team and go through their own performance and look for elements to improve on.

THE CATCH-22: ARE WE FEEDING THE SYSTEM, OR IS THE SYSTEM HELPING US?

'There's no real mathematical validity in a total recordable injury frequency rate, and it doesn't tell you whether your HSE management is effective,' Keith Johnson, Safety Manager in construction, tells me.

'And there have been so many examples of companies having a high or low lost time injuries (LTI), and then something bad happens. Deepwater Horizon is an example of that; Piper Alpha would be another one,' he continues.

'We get into this mental state where we think, "*No, everything's good. We haven't had an incident for so long; nothing will happen.*" If we investigate the theory of a high-reliability organization, we would live in a state of chronic unease.'

Keith has previously worked in various sectors in Australia, including mining, manufacturing, and agriculture, and for the last couple of years, he has been in construction. In construction, he works at what he calls 'the asphalt side of things': putting the labor and product on the ground. In terms of incidents, they deal with traffic-related scenarios such as speeding traffic of drunk drivers, fatigue from night-shift work, working with hazardous materials (dust, fumes, hot products), risks in certain manual tasks (vibration of machinery), and competency of workers (ensuring the people with the proper training are doing the job).

Does the system help us do our work, or are we working for the system?

– Keith Johnson

Their safety performance indicators are typically first aid cases, medically treated injuries, and lost time injuries. He describes their lead indicators as auditing, safety observations, inspections, work task observations, environmental factors, number of employees trained, and licenses issued.

'I wouldn't think that most of the indicators are helpful. We get a bit lost in the numbers sometimes, which we chase, whereas we probably need to do something more pragmatic,' Keith explains.

For example, he would go to a site and notice that they've done 20 plant inspections in one month, but there wouldn't be any findings written down following the inspection. He would go to the plant manager and say they've just completed one inspection for each day. Keith would ask if they consequently did anything, to which the plant manager would say, 'No, I've just put a number into the system because that's what you want from us.'

'Well, no, that's not what I want,' Keith says.

That is the catch-22: Does the system help us do our work, or are we working for the system?

'Then there's the corrective actions after incidents,' Keith explains.

In one example, the organization decided to put covers over the top of the conveyor belt to avoid dust coming out. He would return a month later and check whether they did it. Yes, they would've done it for one, but then the question arises, did it help them? Is it practical, or is it just creating another hazard? Sometimes precisely, that is what occurs.

'You might find that because they've got this dust accumulator over the top of the conveyor, the dust actually can't get out. And then you create this environment where you could have spontaneous combustion because of overheating,' Keith says.

'The things you can do around due diligence to check, and we're checking the checkers pretty much. Who polices the police? I suppose that's what we're trying to achieve,' he says.

A Litigious Society: Who Do We Keep in Mind When Designing the System?

A few years ago, in Queensland, an employee was unloading some large steel girders off the back of a truck tray, and they fell on him. The organization reviewed the risk assessment, which listed all the steps of the job; the organization noticed that they had a risk assessment in place for the task and that it was fit for the purpose of passing an audit. However, it was not helpful to the end user or the person on the ground regarding their understanding of what was required to do the job.

'That probably just comes back to us being such a litigious society. We're so concerned about what's going on behind the scenes that we cater a document that will pass an audit, but we don't do anything to help the end user. And the end-user was the guy that was unloading it. He needs to know what's happening when he's doing his job and that people have thought about *him* rather than legal documentation,' Keith says.

The Subconscious and Safety Behavior

Keith explains that he started to look into the psychology of the conscious, the difference between the unconscious, conscious, and the subconscious. Safety behaviors seem to be embedded in the subconscious, which is where habits are nested. Apparently, it takes us six weeks to learn and then it's embedded in our heads.

How can this be placed into the safety context?

'You might go to do a walkaround of your truck, and every day, you'll just jump out of the cab, and then you will go clockwise around the vehicle and check everything. We might say, "You should go anti-clockwise to do your checks." However, you've gained the subconscious habit of going clockwise each day. Unlearning a habit is more difficult than learning a new habit. If you're concerned with safety behavior, these are things to consider, but they are often overlooked,' Keith tells me.

There is more to the human mind in relation to safety that we overlook.

'I read this book by Clive Lloyd, in which he explains that the key safety performance indicator is trust,' Keith continues.

Instead of calling someone a safety officer, they can become a safety advisor. Removing a policing title away can create trust. Instead of being an accident investigation, it might have been a learning opportunity or a learning review.

'Terminology can make all the difference,' Keith says.

INTERVENTIONS: LOCAL VERSUS GLOBAL

In Australia, the rail sector uses metrics that have been inherited because it is a highly regulated industry. I spoke to a Senior Human Factors Advisor for the Australian rail sector – and he describes most of them as being 'functionally useless to safety.' Organizations are obligated to collect data that they already know it's not necessarily going to be helpful.

When organizations work to improve safety for specific areas of the business, they find, in general, that local interventions are better than global interventions. In the Human Factors Advisor's experience, large corporate interventions tend to be relatively fruitless, as they generally don't have a significant impact. 'Grass root interventions' work best in his experience: talking to the people who do the work. They identify what's wrong and then the organization facilitates the resources. Ideally, you try to identify regular work rather than follow up on particular incidents. The Senior Human Factors Advisor I spoke to illustrates these notions with a few stories of his own.

> In 95% of the cases, the narrative was that the person who 'inherited' the incident was at fault, and interventions were the employee being stood down or undergoing a drug test.
>
> *– Senior Human Factors Advisor in the Australian rail sector*

The first one is when the Senior Human Factors Advisor was first hired; the company had quite a few 'high potential incidents' – meaning any incident that can lead to a score of 4 or 5 in their risk matrix, which is a single fatality or multiple fatalities, avoided by any unplanned control mechanism. The organization asked him to look at those investigations when he first started. He noted that his circle of influence was, at that point, still very narrow. He wanted to develop a way to communicate what they saw as significant issues in a way that the executives could understand. He and his team took all the findings and recommendations, made a scale from one to 10, and described whether it was a local or global intervention, short-term or long-term intervention, etc.

In 95% of the cases, the narrative was that the person who 'inherited' the incident was at fault, and interventions were being stood down or undergoing a drug test. The organization wasn't learning much from this data, and therefore, the recommendations didn't enable any safety improvement.

Then, a new concept of safety was introduced to managers and team supervisors, like systems thinking and models like AcciMap. Afterward, they introduced these theories to operational personnel, highlighting that human error is not the endpoint but a *starting* point in any investigation. The trick to successful learning is to introduce all concepts at their level.

'The process of changing this mindset is slow because the regulator accepts certain outcomes. If a train driver was distracted and caused an incident, we can say that the train driver received more training and shouldn't be distracted anymore, and the regulator would accept it,' the Senior Human Factors Advisor explains.

DEFINING THE GOALS OF INCIDENT INVESTIGATION

According to the Senior Human Factors Advisor in the Australian rail sector, there are three reasons for investigating: prosecution, learning, and avoiding similar issues. The problem with investigating for prosecution is that during the investigation, you'll almost always find that someone was at fault for not following procedures or for a purposeful violation. Once this is known, it won't account for their workarounds to meet potentially conflicting demands. This usually doesn't fit the mission of the public prosecutor, and the learning potential is lost (see also Chapter 2).

Managing change is done by having the Senior Human Factors people – under the Safety, Risk & Assurance department – operate as an 'internal consultancy.' Colleagues come by with all kinds of human factors-related problems, much of it being physical ergonomics, but more and more moving toward cognitive and organizational issues. Their team's value is to help people solve their problems. They establish their colleagues' trust by delivering and not holding a separate agenda.

> There is a culture in which it is assumed that when someone works in the industry for a long time, they must know what's up.
> – *Senior Human Factors Advisor in the Australian rail sector*

At the beginning of his career, the Senior Human Factors Advisor had never worked in rail before. What he's found is that when meeting new people, the first thing that is always mentioned is how long they've worked in rail. There is a culture in which it is assumed that when someone works in the industry for a long time, they must know what's up. However, their understanding of safety in rail might not be the equivalent of those years in the industry. It might even be zero. People might only just be beginning to understand what's happening safety-wise in the industry.

> Whether you've worked in rail for 20 years or 1 year: we're all babies in terms of understanding what's going on safety-wise. So, don't be upset if we try something that doesn't work.
> – *Senior Human Factors Advisor in the Australian rail sector*

'I *know* that I don't know,' the Senior Human Factors Advisor continues.

'Historically, safety-people want to be seen as "knowing stuff." Knowing how to get safer, how to get trendlines down in dashboards, and which interventions work. But most people *don't know* that they don't know. Because I know I don't know, my role in the organization is less about intervention and more about sense-making. It's about going out with people who do work, observing it, discussing it, having them create the knowledge about what's safe in their environment, what they feel threatened by, and then resolving that for them. Going out, having toolbox talks, and telling people to work safer is not only a complete waste of time but condescending.'

As an example of a concept that doesn't work for him, the Senior Human Factors Advisor in rail mentions the concept of Safety-II.

'Yes, it's very logical when you think about it on paper, but in reality, it's tough to devote the same attention that is devoted to things going wrong, to looking at things going well,' he says.

'Naturally, instinctively, the focus is on things that go wrong. The organization expects safety professionals to look at things that go wrong. This is not to say that through safety performance indicators, you can't put in place a monitoring system that detects changes. Not necessarily things that go wrong, but to detect *changes*. And this includes improvements in a safety performance indicator, so you ask, "*Why is this suddenly improving'?*"' he explains.

SAFETY-I AND SAFETY-II AND RESILIENCE ENGINEERING

When discussing safety, it's often defined as its opposite – the absence of safety. The traditional concept of safety, known as Safety-I, is defined by the absence of accidents and incidents or 'freedom from unacceptable risk.' As a result, safety research and management have typically focused on identifying unsafe system operations rather than safe operations. However, Resilience Engineering takes a different approach, acknowledging that both successes and failures can occur for the same reasons. This perspective, Safety-II, defines safety as the ability to perform effectively under varying conditions. Therefore, to fully understand an organization's safety performance, it's essential to have a comprehensive understanding of its everyday operations (Hollnagel, 2014).

FINDING LOCAL RATIONALITY IN THE DEVELOPMENT OF THE EVENT

Rob Hoitsma – A Human Factors & Safety Culture Specialist – remembers the following event: In 2014, a train accident occurred in Amsterdam where one person died, and several were injured. The train driver disregarded a red light, causing a head-on collision with another train. The investigation discovered that the train driver was on a phone call during the accident, and the public prosecutor quickly determined who was responsible.

It was discovered that the train driver's conversation on the phone was actually a part of their job, as they were speaking with the train traffic control unit. To better understand the situation, it is crucial to bring the prosecutor along to see how the job is performed, to understand their local rationality as to why it was logical for them to do so. While receiving a fine may seem like a simple matter of paying a fee, it does not necessarily make the world a safer place. It is commonly accepted that in situations where someone is at fault, they or their organization must pay for it. However, this approach is not ideal for learning from incidents. If the public prosecutor starts investigating, it can set the organization back significantly. The only control possible is through the internal process of organizational learning, and it is vital to bring the prosecutor on board with this line of thinking. The focus should be on understanding why a particular action was deemed logical.

It's not about safety; it's about how you treat people.

– Rob Hoitsma

LOCAL RATIONALITY

The decision-making process is complicated due to various factors such as people, objects, objectives, regulations, principles, and knowledge. Decisions are made with the limited information available to the decision-maker at the time, in a particular context, and based on their understanding of what is reasonable within that context. Local rationality, where decisions are considered rational by the decision-maker because they are based on the information available in their immediate surroundings and at a specific moment (Eurocontrol, 2014).

Many organizations have a safety department, but in reality, this is a single-issue approach. Safety should be at the core of all other departments, from HR to Operations and every department in between.

DEFINE THE LEARNING FROM INCIDENTS PROCESS

David Van Valkenburg – Safety professional and incident investigator at Humans@ Work – began his career at the technical department of the Dutch Railways as an incident investigator. However, he found that the investigations followed a traditional approach using the simple incident method (SIM). This method was often used to investigate incidents where a train driver had ignored a red light. After the investigation was conducted, a report was written and discussed with the relevant manager, but there was no further learning from the incident.

David believed that the investigation method needed to change, and he wanted to introduce a more visual approach to exploring incidents. He also wanted to involve the people who were involved in the incident in the investigation, which he called a 'learning session.'

In one of the cases, there were four people involved in an incident, and after an investigation was concluded, the directors remained with questions, so they asked David to talk to them.

The investigations team expanded the research and included a multi-actor timeline for the various involved people and made a nice schematic overview of the event. David noticed that one of the involved still felt an incredible amount of guilt, even three months after the incident. That person said, '*I am to blame for what happened.*' But when they started to discuss the events, they all realized that the situation was much more nuanced and complex than they thought.

The sessions with the involved people are one of the most important developments the investigation department has made. The core of the session was the timeline, and behind the timeline, there would be an analysis of events that have happened. The results would then be presented to the involved parties, and they would usually lean back and nod. After a while, the investigation method changed slightly. Now, the investigators would only make a timeline, and then, they would invite the involved people and ask them, 'Okay, tell us what happened,' as they are the ones who go through this process of work-as-done every day. The investigators did this to better understand normal day-to-day work, the operator's world, and their thought processes.

One of the many benefits of this method is that it immediately shows the process and how the events have evolved, rather than a written description which is usually included in an investigation report and more difficult to follow or make a mental representation of.

David believes that the introduction of the multi-actor timeline had a positive impact on learning from incidents within the organization. During these sessions, operators were able to learn from their mistakes and improve their work processes. Occasionally, the director was invited to these sessions to observe the work being done and better understand why incidents occur. As a result, the organization gained new insights into how work should be done and updated its instructions, manuals, and procedures accordingly.

In one example, the organization would see many incidents around manual switches, and the investigators found out that there were three types of switches. All the switches have pros and cons, but the train drivers wouldn't know the cons. The train drivers would only be told how to operate them. The investigators bundled the best practice methods among the train drivers. After this, the organization saw a very sharp drop in incidents around manual switches.

Then, after three years, the incidents with manual switches started to reappear. Many investigations are heavily focused on the investigation methods and conducting the investigation. However, when the results are drawn, that's when the organizational learning starts, the learning from incidents process.

Currently, there are no available parameters to measure the success of Safety-II. David suggests exploring potential (in)direct correlations with other parameters like organizational commitment, happiness at work, and psychological safety.

> People aren't working to be safe; they are working to get a job done.
>
> *– David van Valkenburg*

BEGIN WITH THE END IN MIND

Dr. Ilias Panagopoulos – Air Safety Manager in the airline industry – explains:

> Start at the end of the risk and work back towards what leads to the risk, and if there is no issue there, then you don't have to use it as a safety performance indicator.

For example, if an airline has 20 unstable approaches during a specific period, but a go-around was conducted in all cases, then the airline doesn't have to classify *unstable approach* as a risk. If it has 20 unstable approaches but only five of them result in a go-around, then you might need to consider the risk of an unstable approach.

Indicators need to be coupled with other metrics to say something meaningful. In the next step, one could consider coupling SPIs as to why they happened. For example, why do crews continue unstable approaches? It could be fatigue; it could be time pressure; it could be anything. Once this becomes known, it becomes clear what to control and how to minimize variations in the operation.

His advice: keep collecting the data, look for an average, and then determine the deviations. Deviations can be controlled.

CONCLUSION

This chapter explores the use of safety performance indicators, their benefits, and limitations in organizational safety management. Real-life stories from this chapter underscores the need for a balanced approach that uses safety performance indicators as tools within a more holistic safety strategy. The chapter cautions against over-reliance on these indicators as they can lead to unintended consequences such as gaming the system and overlooking complex safety dynamics. In conclusion:

1. Safety performance indicators are valuable tools for tracking trends and allocating resources.
2. Over-reliance on safety performance indicators can lead to unintended consequences.
3. A balanced approach to safety considers factors beyond mere metrics.
4. Safety performance indicators should complement a comprehensive safety strategy.

REFERENCES

Eurocontrol. (2014). Systems Thinking for Safety: Ten Principles a White Paper. Retrieved from: https://skybrary.aero/sites/default/files/bookshelf/2882.pdf

Hollnagel, E. (2012). *FRAM: The functional resonance analysis method*. Boca Raton, FL: CRC Press.

Hollnagel, E. (2014). The construction of Safety-II. In *Safety-I and safety-II: The past and future of safety management* (pp. 125–145). Ashgate Publishing Limited.

Nederlandse Arbeidsinspectie. (2021). Jaarverslag 2021. Retrieved from: https://www.nlar-beidsinspectie.nl/publicaties/jaarverslagen/2022/05/09/jaarverslag-2021

Roelen, A., & Papanikou, M. (2020). *Measuring safety in aviation: Developing metrics for safety management systems*. Amsterdam: Aviation Academy, Amsterdam University of Applied Sciences.

Safetydifferently.com. (2023). Retrieved from: https://www.safetydifferently.com/wp-content/uploads/2018/12/RestorativeJustCultureChecklist-1.pdf

The Guardian. (2021). Revealed: 6,500 migrant workers have died in Qatar since World Cup awarded. Retrieved from: https://www.theguardian.com/global-development/2021/feb/23/revealed-migrant-worker-deaths-qatar-fifa-world-cup-2022

Vanraes, B. (2020). A Dark Secret of Safety. Retrieved from: https://www.linkedin.com/pulse/dark-secret-safety-bart-vanraes/?trackingId=0RFVAppcSvOdrNade2qSBA%3D%3D

5 First Rule of Leadership
Everything Is Your Fault

Leadership is inspiring people; management is keeping the trains running on time.

– Andy Dunn

ACCIDENTALLY EFFECTIVE SAFETY LEADERSHIP

Leaders and executives are responsible for making sure that they understand what drives and motivates their workforce. What might work in one team does not guarantee success in another team. Effective safety leadership requires a combination of skills, such as effective communication strategies and consistent messaging about the importance of workplace safety at all levels within the organizational structure. And, of course, incident investigation to keep learning from adverse events. Because if the organization does not understand why incidents happen, how will it keep learning and improving? Understanding why something goes wrong, and also how the system can better support people in doing their jobs consistently well, offers critical insight into how the organization can keep learning and improving.

Carsten Busch – whom we met in Chapter 3 – tells me a story in which not all of the above leadership strategies were applied, but it weirdly turned out well. Unfortunately, the trigger was a train accident. At the time, he was working at a Norwegian rail infrastructure agency.

The Sjursøya train accident in 2010 was one of the most significant accidents in Norway. A set of sixteen freight cars began to roll uncontrollably during shunting on Alnabru, north of Oslo. The train sped downhill for several miles. The train hurtled out of control, smashing into a building and plunging into Oslofjord. The structure, where several people had been working, collapsed, and part of the train fell into the fjord. Three people were killed in the accident, while four people were injured. The Norwegian Accident Investigation Board was tasked to investigate the accident.

What did Carsten's boss decide after the accident?

To *not* allow its safety staff to conduct an internal accident investigation.

'At first, there was utter outrage that we were merely to support the national accident board's official investigation. We would rather concentrate on creating some barriers to prevent a recurrence,' Carsten explains.

'Later, as it turns out, this decision turned out to be the uppermost New Safety View that could have happened.'

How so?

Management did not allow Carsten and his team to investigate this accident; they were only allowed to assist the Norwegian Accident Investigation Board in their

DOI: 10.1201/9781003383109-5

investigation. After feeling professionally insulted, the team eventually complied with this request.

Instead of researching the accident, the team focused on conducting a new risk assessment. Many things had changed since the last risk assessment: the organization had changed, the shunting area had changed, procedures had changed, and contractors had gotten involved, all enough reasons to analyze the work as it was now being done. Their goal was to involve the people as much as possible in this assessment, so Carsten and his team assigned a consultant who would spend full-time efforts to get this information. Not only to speak with the different teams but also to join them in their shifts to see it with their own eyes.

Five years after the accident, Carsten realized that the constraint to investigate this accident was the best thing that could have happened.

'If we were to investigate this like we always did, then we would have focused on what had happened and made recommendations based on that. Because we were forced to ignore what had happened, we looked at the context in which this accident could have happened and addressed the whole context. This led not only that the context in which work is being done has become safer, but the work has also become more efficient,' Carsten explains.

After the Norwegian Accident Investigation Board delivered their report, there was nothing new in it for Carsten and his team as they had already been able to evaluate the context in which the accident could have happened.

Did the person who prohibited the team from investigating know what (positive) effect this would have?

'No,' says Carsten.

'It was a purely political decision, which could have been explained better at the time to avoid the team feeling so angry and frustrated.'

'Also, this is a good example of a complex system: it is difficult to predict what effects certain decisions will have,' Carsten says.

SETTING CLEAR EXPECTATIONS FROM LEADERS FOR SAFETY PRACTICES

In recent years, there has been increasing attention to the leadership styles necessary for effective safety management. Today, safety leadership is seen as a critical component of facilitating safety. Leaders have the power to shape the attitudes and behaviors of their employees and can play a crucial role in creating a safe work environment.

Different leadership styles have been studied in terms of workplace safety. Transformational leadership has been found to be very effective in promoting a positive safety culture (Hofmann & Morgeson, 1999). Safety literature provides empirical support for the positive impact of transformational leadership on workplace safety attitudes and behavior (Zohar, 2004) and even organizational performance (Geyer & Steyrer, 1998). Transformational leadership style emphasizes communication, empowerment, and giving employees individual attention. On the other hand, laissez-faire leadership – where leaders are not involved or disconnected – has been linked to higher injury rates (Huang et al., 2015). Research suggests that prioritizing safety as a core value and actively communicating this message to employees can increase engagement with safety practices and improve overall safety performance (Clarke et al., 2010). Training interventions showed that leaders' safety attitudes were highest among managers who received the safety-specific transformational leadership training, as opposed to managers who participated in the general transformational leadership training (Mullen & Kelloway, 2010). Training a small portion of organizational members (managers) has a significant impact on a large number of individuals within the organization. The safety-specific approach to training leaders is a cost-effective and efficient way to move forward in safety management within organizations.

Leaders should set clear expectations for safe behavior and encourage open communication about potential hazards or employee concerns. It's also essential for leaders to be approachable and available so that employees feel comfortable reporting incidents or near-misses without fear of negative consequences (Fleming & Lardner, 2002).

In one study that looked into safety and accident prevention among managers (Rundmo & Hale, 2003), ideal attitudes for managers that were identified are high management safety commitment, low fatalism, low tolerance of rule violations, high worry and emotion, low powerlessness, high safety priority, high mastery, and high risk awareness. High management safety commitment and involvement are particularly important for the managers' intentions and behavior.

Taking these – and other studies – into consideration, it seems that leaders should possess both good leadership skills but also be very 'safety aware.'

But Should All Managers Have Operational Experience?

Not necessarily.

At least not necessarily according to Nuno Aghdassi, an Air Safety Investigator in Portugal and former Head of a leading private jet operator.

'As long as this person possesses safety expertise and brings in advice from the operational staff,' he says.

'That's key.'

Nuno elaborates that it is even more critical to have the right people skills, the so-called soft skills.

'But it's the personality traits which allow you to communicate effectively, to gain people's trust, to work with people, to foster team building, and to use that to improve safety performance. To make the best decisions as a manager, as a result of bringing together the necessary expertise,' Nuno says.

It is up to the organization to decide what they want to focus on. Indeed, in the aviation domain, it is increasingly important to have good decision-makers, people who can manage teams and do so effectively, and in doing so, bring together the necessary expertise required to help them in their decision-making.

What is essential is the way management communicates, especially about safety. People are sensitive to implicit signals. In communicating, there is the explicit message: be aware, be vigilant, be safe, etc. The implicit message is what is said about safety behind closed doors through the examples set by oneself, the unspoken messages management conveys about safety, or people in the safety department.

Whether we like it or not, this is a top-down process. People 'dance to the tune' of the person at the top.

'It starts top-down, and you can talk about safety all day, but if you do not set an example yourself – if you as a chief pilot or as a director of flight operations – if you do not submit that safety report when it is needed, or if you do not follow the Standard Operating Procedures in an abnormal or emergency situations, or if you discredit someone in the safety department; it's not gonna fly,' Nuno says.

PRIORITIZING SAFETY ALONGSIDE OPERATIONAL AND COMMERCIAL GOALS

According to Nuno, changing the hearts and minds of employees should occur not necessarily at the top because the support of senior management is usually ensured. In general, the challenge of change comes from the middle managers. Some of the directors, but generally the managers who report to those directors, are the ones who are the challenges because what happens at that level is they know people talk about safety. They know that safety is a priority for them. However, they do not feel the same level of accountability as an accountable manager or director. Additionally, they have many conflicting goals: they have operational goals, they have commercial goals, and they have safety goals.

'Because the weight of accountability is not the same, signals get mixed up typically at middle management. You will find that organizations typically have the greatest challenges with those middle managers. They will talk about safety, they would have heard about safety, but will they ever sit down and talk about safety with their teams? Rarely will they do that. They will sit and talk about procedures, commercial goals they must meet, and their targets in terms of aircraft movements, for example,' Nuno describes.

Middle managers' bonuses are typically associated with commercial and operational performance but not safety (unless you work for an organization that does, see Chapter 4). Their job descriptions may or may not specifically mention safety as something measurable regarding their performance. That is where the most significant challenges are.

Additionally, some people have been put in a safety management role on a part-time basis.

'I think you need people who have a clear vision or ideas and have been able to transfer the theoretical into practical hands-on actions. They know exactly what to do because that is how you need to explain it to others,' Nuno elaborates.

'If you go to your accountable manager or your directors, and talk about the theory, it's going to sound mumbo jumbo to them,' he illustrates.

'If you can translate that into practical things which apply to that organization, which influences the day-to-day work of others, then I think that gives it different credibility. Credibility is the key to trustworthiness. You need this because the safety department is where you need to gain the trust and support of front-line staff,' Nuno continues.

ALIGNING ALL MANAGEMENT LEVELS WITH THE SAFETY GOALS

At the same time, Nuno finds that you need to gain the trust and support of senior managers. This is a fragile line to walk because you do not control everything that happens. You do not control what the directors will say to the pilots. You do not control the behavior of the employees. You can have all the technical tools; you can have a safety reporting system; you can have all the procedures. However, it takes much time to build that trust and build people's confidence in becoming involved and engaged in safety. Very quickly, that can become undermined by something outside your control.

For example, communication that goes from the director of flight operations to the pilots, which may not be worded correctly or may have a word that is misinterpreted, can already put people in a defensive mode. Therefore, it is helpful to have an aviation psychologist work full-time in the safety department because every aspect of what is done, other than technical or operational, is psychological.

EMERGENT BEHAVIOR FROM SOCIETAL DEVELOPMENTS

At the leading private jet operator where Nuno worked, they were monitoring safety reporting performance indicators. They would find quite clearly that, in times when there was a recession, those times when people felt less secure about their jobs, or when a communiqué went out from the president of the company, there would be a drop in the number of safety reports, and this had nothing to do with the operation. It was the sentiment determined by people's level of engagement with the organization. This example serves to show that there are some forces at play outside the organizational control that may have an effect on people's behavior, including safety behaviors such as safety reporting.

SERVANT LEADERSHIP: EMPOWERING EMPLOYEES WITH GROWTH AND JUST CULTURE

In line with what Nuno has explained, Martijn Flinterman – a Senior Advisor Integral Safety at a public infrastructure management department – urges all organizations and management to reassess how they communicate about incidents – both internally as well as externally.

'It is understandable that it's in human nature to consider who is the party to blame when mistakes are made, such as the road users after an accident. Spokespersons will then state that it's the fault of the road users and that they should pay more attention or comply with the rules. However, it is our responsibility as a governmental body to build forgiving roads to enable road users to correct their mistakes rather than building roads that make them drive into a tree. And then you get a news headline stating that road users do not comply with rules. This is frustrating because it does not invite any parties to have meaningful conversations about what safety means. Let's not focus on how the organization looks to the outside world. We can only really manage what we can influence, and the rest we need to accept.'

Martijn started as a sociologist 20 years ago at a European public body responsible for infrastructure and waterways, roadworks and construction, bridge safety, and road safety. He initially worked on how safety and risks are experienced and looked into their safety policies.

In Martijn's project teams, several safety experts are involved: Health & Safety coordinators, explosives experts, human factors experts, construction safety experts, road safety experts, etc. They appeal to their expertise when needed to enable multiple views on safety during various phases of a project.

At a certain point, the organization became more market-focused, and 'then you get the Anglo-Saxon measurable and performance indicators for free,' Martijn says. Of course, nothing is wrong with that, but it could come along with undesirable effects, such as underreporting to keep the level of measured incidents down.

One of the key focus points for safety within their projects is whether safety has been integrally evaluated. This starts at the procurement phase, continues with project budgeting, and throughout the project. The idea is that they avoid looking in hindsight whether there is still time and resources to consider safety aspects. The organization wants to step away from just looking at its numbers and keep looking at the dynamics in which safety is balanced.

'We usually use the dichotomy of what is considered safe and unsafe. To me, that doesn't make any sense. I much prefer to use the terms risk and hazards. Risks need to be weighed internally, while hazards are external factors the organization must deal with. In organizations, the question of "*Is this a safe way to operate?*" is insufficient because nothing is completely safe or unsafe. It is about the weighted and relative safety, which needs to be included in all projects,' Martijn says.

Moreover, during his long tenure at the organization, he has experienced the necessary frustration of running the daily business. He decided to talk to his manager and explain that to develop as a safety professional, he needs time and space to conduct his own research, explore other methods, have some side projects, and do experiments now and again.

Luckily, Martijn has received the support he asked for, but the question is how to safeguard this space from being run over by the daily operation. Historically, not all of his managers were open to this idea. However, recently the organization started to select its leaders to fit the 'servant leadership,' which focuses on service to employees, customers, and the community. Whereas traditional leadership is often concerned with power, status, and hierarchy, servant leadership focuses on employee awareness, meaning, development, and growth.

> If you're used to thinking in legal terminology – which is inherently compliance based – it's a tough pill to swallow when a report summarizes points of improvement for the organization: You're thinking about the organization's liability.
>
> *– Martijn Flinterman*

In the past, the organization has had a functional resonance analysis method (FRAM) made for the topic of roadwork safety. Martijn says this topic is typically something that everyone within the organization points to rules and procedures, the compliance approach of following those in all cases. However, considering the daily operation, many factors are at play, such as whether the roadworkers have enough time to conduct their work. Especially when this type of work is outsourced to external contractors, they are inclined to stick to a tight deadline because the 'delay fines' are hefty: Martijn mentions €50,000 for a 20-minute delay of not having a road opened.

Logically, those companies doing the job will do whatever they can to avoid such penalties. Consequently, stakes get intertwined, and decisions become biased: leaving specific equipment as is, crossing a busy road quickly to get some road signs away, finishing a job fast rather than safely. It becomes pointless to focus on compliance as an auditor. The context in which these 'unsafe behaviors' occur is simply the result of the balance between all stakes. Instead, the focus should be on having a meaningful conversation between operators and auditors, policymakers, procurement, and contractors.

These insights inevitably lead to unease and resistance at the management level: Operators are knowingly and willingly exposed to hazards and risks. It is much easier for management to say, '*We have rules and procedures in place, and if operators do not comply, it is their fault.*' More straightforward, for sure, but it does not reflect the complexity of the work. However, if you are used to thinking in legal terminology – which is inherently compliance based – it is a tough pill to swallow when a report summarizes points of improvement for the organization: You're thinking about the organization's liability.

At the same time, management is confronted with various stakes as well. The list may become very long from managing finance, rosters, and all relevant internal and external stakeholders. Safety is only one part of many other stakes, and getting the safety message across can become challenging. It is also easier to blame management for not hearing the safety message, but at the end of the day, aren't they also 'humans at the sharp end,' just like the operators? Wouldn't we all benefit from walking a mile in each other's shoes? We all know the CEO that spends the day in the mud, but what about the operator that spends the day in the CEO's office?

DESIGNING OUT RISKS: SETTING CHALLENGING OPERATIONAL GOALS

'Figure out what you want and then go and build it,' Gretchen Haskins says, Board Director for HeliOffshore and the Flight Safety Foundation. For her, the motivation for improving safety came from a very personal space, in which she had lost a loved one at a young age in an accident. She did not want others to lose their loved ones, too, so she became interested in safety, but mainly in the tangible *results* of safety. In her opinion, to get observable results, operational goals and requirements need to be set, ideally from the beginning of a design phase. However, she has plenty of examples that show how setting goals and requirements – even after the operation is already running – can ensure safety success.

HeliOffshore is a company dedicated to global offshore helicopter safety. Gretchen is an aviation industry leader in safety performance improvement and an internationally recognized expert in human factors. She has served on the UK Civil Aviation Authority board as Group Director of Safety, overseeing aviation safety in the UK (including airlines, aerodromes, air traffic, airworthiness, and personnel). She also worked as the Group Director of Safety at the UK air traffic organization National Air Traffic Services (NATS), where she championed activities to support frontline operational safety.

Safety shouldn't be a check function. It should be a design function.

– *Gretchen Haskins*

ASSESSING RISKS IN A LIVE OPERATION

During her time at NATS, the safety team analyzed around 6000 reported events yearly. Most of these events were not serious. They had a good and functioning safety management system in place, and they never experienced a fatality as a result of providing air traffic services. Because they never had a midair collision, one could say that they were 'pretty safe.' However, at some point, they changed their approach to preventing a midair collision over London. They moved from managing the risks associated with just their operation, toward looking at all causes, even if it was not their fault. They understood that an event like this would be beyond catastrophic, and they started to dig beneath the surface to see what more they could do to prevent it.

One thing the organization looked deeper at was what causes there were for these events. They devised 'the usual,' such as airspace infringements, level busts, and runway incursions. Then, they asked what we must *do on a typical day to keep us safe and prevent those events.* They invited the CEO to join in this conversation. He said he wanted more than the usual safety goal of 'let's make it as safe as possible and no worse than before as traffic grows,' which is common in air traffic management when changing the system. Instead, he said, '*Let's set the goal of designing out our most serious events.*'

One of the first steps was to align around the top events that could result in a fatal midair or ground collision, no matter the cause. They then started discussing this with various stakeholders, such as their customers (airlines), the regulator, the board, the frontline personnel, and others. Everybody came up with different stuff, but in the end, they had an agreed list of top priorities: airspace infringement, level busts, and runway incursions. Airspace infringements are situations in which any aircraft enters controlled airspace without permission. Level busts are situations where an aircraft is cleared to climb or descend to a certain flight level, but the aircraft overshoots the permitted altitude. Runway incursions are occurrences in which there is an incorrect presence of aircraft, vehicles, or persons in the protected area designated for takeoffs and landings.

A forward-looking strategy to design out those risks was the result. First, they set operational goals; for example, for airspace infringements, they identified air and ground functions that need to be performed consistently well to prevent the risk. These included flight planning, navigation, detecting, avoiding, and/or resolving the situation. They looked at potential ideas to enhance performance against the incidents that had happened and what percentage of the previous incidents they thought the intervention would have stopped. They asked themselves, would it have stopped it at 100%, or would it improve by 50%? They created a model that said, '*We think we can improve this much if we do this technology or this training, or this intervention.*'

What they started doing was making business cases for safety. They defined that every design or project, whether procedural or technical, should have safety benefits described and the long-term investment plan for the organization needed to be explicit about the resources put into achieving this benefit.

For each project, they set safety goals. Improving detection should be done by some time in the future, faster, more accurate, timelier, etc. They did not eliminate 100% of their risk, but they designed out two-thirds in three years. The result was that they stopped having two-thirds of the most significant events they previously had.

Similarly, they looked at the number of level busts within their operation. They wondered how much of a problem it was, knowing there are several hundred each year. The airlines held much data on level busts, so they approached them and shared their data. Together, they realized that the problem was much bigger than everyone thought. They started to look for causal factors and found that one of them was that the pilots read back the wrong altitude. At NATS, they decided to watch the Controller's work and asked the controllers to write down every time ATC noticed the pilot read the wrong altitude and corrected it. Controllers wrote down as many as 1100 wrong readbacks in 10 days! Then, they got tired of writing it down.

The project team checked how often the Controller missed a wrong altitude readback. They were about 99.9% accurate. Great performance but it meant that once every ten days, the Controller or the pilot read the call back wrong, and the Controller missed it. As there are 7000 flights over London daily, at least once every ten days, they had an airplane at the wrong level, and nobody would know.

That insight led to NATS wanting to further improve detection. They used this investigation to make the business case for a small amount of software that downlinked to the Controller's selected flight level and gave the Controller an alert if that were to deviate from the instruction. This significantly improved performance. 'And it all started with the higher-level goal of getting rid of the potential for a fatal mid-air or ground collision no matter the cause,' Gretchen clarifies.

Doing the Same = Getting the Same Results

With 6000 reported events to be investigated each year, NATS would spend millions of Pounds on those investigations and millions of Pounds on implementing the recommendations they kept making over and over. At some point, NATS agreed with the regulator not to investigate the same events repeatedly if they were similar to an event investigated earlier. It would free up their resources to spend more time on better solutions to the existing issues and recommendations.

NATS introduced a feedback loop in which they described what benefit in terms of performance they expected to see after the recommendation was made and kept track of whether the recommendation had the desired effect, often in better supporting frontline performance.

Because of this strategy, the investigation team did not need to come up with as many recommendations as before. They were now tracking the progress of the running recommendations, which lead to increased safety benefit.

Aligning Personnel through Communication: Heads, Hearts, and Hands

At HeliOffshore, they knew a significant proportion of fatal accidents were controlled flight into obstacles or surface, or terrain, also known as CFIT. They decided that they wanted better obstacle detection, knowing that the airlines had put in a better ground proximity warning system and virtually designed out CFIT for the aircraft.

There were over 1000 fatalities worldwide in the decades before the airlines found a solution, and there were zero after the airlines installed the new system in their aircraft. For helicopters, they also wanted a terrain avoidance system to function this way.

However, as it turned out, the algorithms in the system were not attuned to how helicopters flew offshore. Those algorithms were from commercial jets, and their trajectories and speeds were not optimized for offshore helicopters.

A research project was initiated to define some new operating envelopes for the algorithm and test it in the simulator. They tested a set of previous fatal accident scenarios, and in all cases, the pilots were able to detect and prevent the event.

Meetings were organized with the executives of helicopter operators, manufacturers, and customers, and they explained why they wanted to pursue this. HeliOffshore showed the data, and everyone thought it was a no-brainer. They all set the goal to have this system available in two years and wrote a specification that anyone could use. Rather than any one company, everyone in the world could design it if they wanted to.

However, although the CEOs of the manufacturers had agreed, the engineers were not enrolled in the project. It took the organization a while to realize they were not doing what they said they would. They had to re-enroll, listen to their concerns, and ensure the proposed solution addressed them.

One concern was that the scope of the project needed to be reduced. They had to take one of the modes out of the project that was too contentious in order to get the other six modes through quickly.

'What I have mostly found,' Gretchen says, 'Is when things are not working, it is a communication problem. Moreover, if people are not taking action, their heads, heart, or hands are not engaged. You need to find a way to inspire their head, heart, and hands so they connect with why they want to do it. They can logically see that it needs to be done and know what to do. Almost always, that is the problem. Sometimes it is because they know that the solution they are putting in is not good enough. They have tried to tell everyone to watch out or be careful. They need more resources to put in a better barrier rather than relying on it. Whatever barrier they have is not strong enough.'

PATIENCE, PERSISTENCE, POSITIVITY: KEYS TO EFFECTING CHANGE IN ORGANIZATIONS

Proponents of change do not necessarily see themselves as rebels, according to Dr. Nicklas Dahlstrom – a Senior Lecturer, Lund University School of Aviation. Ideas can be rebellious, though, challenging the status quo. It is the way of *implementation* of that idea that can cause pushback. To successfully implement change in large organizations, you need the three 'Ps,' according to Nicklas:

1. Patience,
2. Persistence, and
3. Positivity.

On patience, he says that you must have many conversations; while you do that, you will think that nothing is happening. Ideas need to mature, and that is often underestimated. So, it would be best if this is done persistently.

'Don't forget that you also must talk about what is going well and what we can improve because if you only talk about what goes wrong, people get tired quickly,' he says.

> Don't forget that you also must talk about the things that are going well and what we can improve because if you only talk about what goes wrong, people get tired quickly.

> *– Nicklas Dahlstrom*

LOOKING BEYOND THE KNOWN: CROSS-INDUSTRY INSIGHTS AND GAINING FRESH PERSPECTIVES ON SAFETY

When Zoë Nation – a Human & Organizational Performance Lead in Australia – was brought in by an oil and gas company to review the safety management system; she and her team looked at several things. They looked at their organization internally and at other organizations, including outside their industry. They held more than twenty interviews to see their steps to improve safety. As a result, they were able to perform a gap analysis. They weren't necessarily looking for a new way of thinking, as the operations department is prone to become safety-fatigued.

The first thing she and her team did was delete the studies with no added value. The organization started working with the investigations team and gave them some knowledge about the fundamental causes of human error, just culture, more Safety-II topics, and how to build trust and create psychological safety among employees. Then, they introduced learning teams. That got some backlash from those who worked in the generally conservative company. They wanted to see more complex data. The department started by discussing incidents in Learning Teams or how the day went in general. It was difficult for the conservative seniors to comprehend how someone who had not followed the rules could 'get away with it.' The seniors considered that the person would want to save his skin. Zoë and her team said those frontline operators would share even more information if the organization reassured those employees.

What they saw was a change in management's approach to failure. The organization no longer went after employees to correct them. The categories of their root-causes changed, so they learned more from their incident investigation. The corrective actions were drawn up for higher management layers of the organization. Bonuses were no longer linked to minor incidents. They gave more autonomy to the local operator to decide what to focus on and started to welcome things like 'bad news' and sharing incidents.

Eventually, the organization started to see a significant change in their corporate culture, which also became visible in surveys. One of the metrics Zoë heard being used by what she describes as a more progressive organization was the 'number of projects solely concerned with improving the control aspect of safety-critical activities.' This demonstrates a proactive attitude to improve safety rather than other audit-like metrics, such as closing all pending incident actions. Leadership commitment is about how management responds to incidents.

REBRANDING SAFETY: ENGAGING EMPLOYEES
IN A NEW NARRATIVE

Similarly, Kym Bancroft – Head of Health & Safety in Australia – and her team embarked on an ambitious project to transform safety and the safety approach. A crucial element of this approach was to shift from seeing people as the problem to seeing them as part of the solution. Learning from what goes right and viewing safety as an ethical responsibility was also part of a paradigm shift in how the organization viewed safety.

Kym's team found that the organization needed to involve everyone when implementing their safety strategy to succeed. To do this, Kym developed a highly successful bespoke training opportunity for Health and Safety Representatives to train frontline leaders and executives.

This training led to recognizing the need to shift from blame toward a just culture and rebrand safety by moving away from zero-harm slogans. This was then supported by an effective metrics system that communicated the desired goals throughout all levels of the organization. The team also introduced the Learning Team methodology as an alternative approach for investigating incidents, aiming to reduce the workload associated with the current investigations while ensuring that every incident was investigated correctly. Compliance is critical for any incident investigation or safety system. However, the process must ensure that people are not afraid to admit their mistakes for learning to occur, thus making sure incidents are not repeated.

The team ensured their efforts were supported by applying change models to operational personnel and 'backend reporting systems.' By involving employees and giving them autonomy, they saw a significant change in their corporate culture, visible through surveys and metrics such as the number of projects dedicated to improving safety-critical activities.

Every organization has its way of addressing safety issues, and Kym's approach to safety proved highly successful. By involving employees, applying change models, rebranding safety, and creating a just culture, she encouraged her team to think differently and learn from each other while complying with health and safety regulations. Her story serves as inspiration for organizations looking to implement adequate safety strategies.

'Just go out and do it. You do need to know the basics and be well-read. Pair it up with the organizational challenges and context. In doing so, be strategic and holistic,' Kym says.

Kym's story exemplifies organizations looking to implement ambitious safety strategies.

'If you are clear on your values about safety, teams need people who challenge the status quo,' she continues.

Aligning safety values should be consistent across all management levels, starting from the top. By ensuring alignment in the thinking about safety, the organization can show vulnerability and underline it's a learning process.

Showing vulnerability works on multiple levels. For example, Jaap van den Berg (Safety Culture Manager at a major European airline) has made a video about just culture with the Chief Operations Officer (COO) of the airline – who also happens to be a pilot within the airline – in which he admits that he has also made mistakes in the past. Mistakes made were discussed and learned from, and he explained how he did not get penalized for them. When other operational personnel see this video, from other pilots to baggage handlers, they could realize that if even the COO can make mistakes and admit them for the more significant cause of learning, everyone should be able to admit them.

CONCLUSION

By following our safety professional's examples, organizations can facilitate a culture of safety that is effective, engaging, and motivating for employees of all levels. By investing in training, involving everyone within the organization, and creating an environment where people feel comfortable admitting their mistakes without fear of punishment, organizations can create an atmosphere where mistakes become learning opportunities rather than punishments. With a combination of health and safety regulations and just culture principles, organizations can learn from each other while complying with regulatory requirements.

In the following chapter, we will delve further into why open communication, good listening skills, and other leadership principles lead to positive changes in organizations and safety.

REFERENCES

Clarke, S., Simpson, J., & Hinder, M. (2010). Leading and managing safety in the workplace. *Safety Science*, 48(2), 168–177.

Fleming, Q., & Lardner Jr., R. (2002). Fixing incident investigation systems: A key component of successful safety management systems. *Journal of Safety Research*, 33(3), 253–266.

Geyer, A. L., & Steyrer, J. M. (1998). Transformational leadership and objective performance in banks. *Applied Psychology*, 47(5), 397–420.

Hofmann, D., & Morgeson, F. (1999). Safety-related behavior as a social exchange: The role of perceived organizational support and leader-member exchange. *Journal of Applied Psychology*, 84(5), 605–614.

Huang, C.-W., Chen, H.-C., & Chiu, S. I. (2015). Leadership engagement and its effects on workplace safety performance: A study of the Singapore construction industry. *Safety Science*, 75, 1–10.

Kelloway, E., St-Jean, É., Francis, L., & Prosser Shaw, K. (2013). Exploring the relationships between servant leadership and employee outcomes in a Canadian federal public service organization. *The Leadership Quarterly*, 24(1), 48–61.

Mullen, J., & Kelloway, K. (2010). Safety leadership: A longitudinal study of the effects of transformational leadership on safety outcomes. *Journal of Occupational and Organizational Psychology*, 82, 253–272. doi: 10.1348/096317908X325313.

Rundmo, T., & Hale, A. R. (2003). Managers' attitudes towards safety and accident prevention. *Safety Science*, 41, 557–574.

Zohar, D. (2004). Climate as a social-cognitive construction of supervisory safety practices: Scripts as proxy of behaviour patterns. *Journal of Applied Psychology*, 89(2), 322–333.

6 Remember
You Have Two Ears

> Most people do not listen with the intent to understand; they listen with the intent to reply.
>
> *– Stephen Covey*

INTRODUCTION

In an operational workplace environment, progression in safety may fall short for various reasons, from psychological safety to resistance to change and everything in between. Industry experts share some strategies for overcoming resistance to change in the workplace. Other concepts and practical advice on creating a culture of innovation are discussed by building trust, fostering collaboration, and promoting open communication. Creating an environment where people are comfortable taking carefully considered risks and learning from mistakes should lead to breakthroughs in understanding each other's work. By understanding the stories behind resistance, businesses can make meaningful progress toward their objectives.

REFRAMING THE CONVERSATION: HOW TO MAKE SAFETY A PRIORITY IN THE WORKPLACE

Meet Ron Gantt, a principal health and safety consultant specializing in construction based in California, United States. He was called in by a large construction company in the Midwest to assist with an issue involving their customer, a healthcare organization that owned a hospital. The hospital was being built amid the COVID pandemic.

Big construction sites typically have a bunch of job trailers where all the project managers and engineers will sit and do their work. And then, there is the area where the construction work is happening, often separated by distance. In this instance, the trailers and the construction site were separated by a big parking lot, and the geographical distance was more than symbolic.

The construction organization was having a problem because the customer said they had too many incidents, and they shut down the job. The incidents were not major incidents; these were minor incidents. Moreover, still, the customer said they needed to shut down and change some things.

In the US, construction assignment owners usually have a hands-off approach to construction. However, this particular customer was more involved and worried about incidents at the construction site. They expected the construction company to find a solution.

The construction company did several things to improve incident rates. One of the things that were particularly effective from early on was rigorous training and a regular safety climate survey. They had pre-job planning and all the usual stuff that one could

DOI: 10.1201/9781003383109-6

expect from a mature company to do. However, one of the things the company did not have was a mechanism to see how their efforts were working – apart from the incident rates. The organization needed to increase its ability to learn about what is happening daily. The organization initiated the use of Learning Teams, something we have seen quite a few times at various organizations throughout this book.

The other initiative they took was the so-called Gemba walks. Ron chose the name because he saw a document on the table where they were meeting as they discussed it. It said the company wanted to start doing Gemba walks as part of their Lean initiative, such as quality management. Gemba is a Japanese word that means 'at the place'; in other words, you go to the place where the work happens to figure out what is happening. The company wanted to get people in that office to talk to workers. Not talking to workers to coach for compliance, saying, *hey, you are not wearing your safety glasses* or something like that, but genuinely asking questions about the work to understand the work.

The organization started a basic communications training program for everybody because regardless of job title or academic success, the company did not automatically assume everyone knew how to go and talk to people. Often, organizational leaders are promoted because of their ability to do technical tasks, not necessarily because of their soft skills. The training was a short two-hour training with everyone that was going to be involved with the Gemba walks showed them how to do it, the kind of questions they could ask, and the purpose of the walk.

Who was doing the Gemba walks? Every leader on the job. The organizational hierarchy had project engineers who were entry-level out-of-university engineers managing parts of the project. Then, there would be superintendents who were field-based. There would be project managers; assistant project managers; and project managers who manage the budget, schedule, and vendor interface. Finally, there was the project executive. The company wanted everybody to go out and do the Gemba walks because the thought was, '*If you are influencing and guiding the work, you should understand the work.*' Employees were initially skeptical, but they were willing to try this.

OPPORTUNITIES FOR IMPROVING INTERPERSONAL COMMUNICATION IN THE WORKPLACE

Essentially, during the Gemba walk, the walker would go up to the person executing the work and ask them, '*Hey, what are you working on today?*'. As they started talking about their work, the Gemba walkers would ask if they had what they needed for that, what kind of tool they used, and if they would ever have any difficulty getting those. Sometimes, they would notice many people working on the site, asking them if they have worked with them before. Questions to understand some of the nuances of the work. Do they have difficulty getting the resources? Is there good cohesion between the crews, and how is that affecting their ability to get the job done? Do they feel bored with the schedule? Do they feel like everything is great?

The organization started this at one job site, and it started working well. The company noticed it worked well because people, like the engineers, were identifying issues they did not know. Ron recalls that they were asking them about it, 'Hey, how is it going? It is just you, I notice. How long does this normally take? Moreover, I see you are on the floor a lot. That has got to hurt your back?'

The worker explained that everything is fine, as long as they are not pushed for schedule because then they could rotate through.

'But if we get pushed for schedule, then we all have to do the work simultaneously, and there is no rotation.' Ron says, 'And as we walked away, I was with the manager for the job, and I said, "what did you learn?"'

The manager said, *'Wow, I did not realize that if we push the schedule on them, that is going to create some ergonomic problems. I was about to push the schedule on them next week!'*. To him, that is an excellent example of a leader now understanding the consequences of choices they would not have known otherwise.

At the same time, production needs to keep going, but it cannot happen by just saying, *'Work harder.'* They realized it was perhaps time to get the frontline staff more resources. They usually would have had more people, but some had to go to a different working position. Now, the manager could ask if there was any way they could get some of those people back. Because then they could push the schedule without compromising people's safety or ergonomic injuries. All the Gemba walk did was help the manager understand the constraints on which she can base a decision. The manager can make a better choice.

CREATING A PROFOUND UNDERSTANDING OF WORK AT ALL LEVELS

The construction company Ron worked for did not have a checklist for the Gemba talk; the talks should be conversational. They provided a list of 'fallback questions' because sometimes someone would go and talk to people, and the conversation was going nowhere. The leaders were not getting anything – some questions that have proven helpful in those situations.

For example, one of the questions is, *'What is something that you wish management knew about your job?'* Another popular question is, *'What do you wish they told you before you started this job, but you had to learn independently?'* *'What is something that you see new people do, which is a common mistake they make when learning?'*. One of the most favorite questions was, *'What is the dumbest thing we ask you to do around here?'*. People love that question.

Another one of those questions involved asking what improvements operational personnel would make if given a sum of, say, $40,000, and they could spend it to make their job better, safer, faster, easier, or whatever. What would they spend it on? It's not about actually giving this money to spend on solutions, but what the question reveals is that people come up with simple, practical, and cheap solutions that work – much more often than expensive mega-changes that management might expect.

What is the dumbest thing we ask you to do around here?

– Ron Gantt

CREATING COMPANY SUCCESS FACTORS EVERYONE BELIEVES IN

Apart from the Gemba walks, another initiative was launched within Ron's organization. They started thinking about creating an assessment process designed to assess how they, as an operating group, are setting up their sites for success rather than grading them in an audit-type manner.

The organization decided to design success factors that they believed would make a project successful. They believed that if they set the project up for success, not only would it lead to a safer job site, but it would also lead to more successful jobs.

A team was brought together from across the operating group, gathered stories about how work is actually happening, and then came back together and asked, 'What do these stories teach us about how we are setting up this site for success?'.

They intentionally brought people from other sites as part of the team from different organizational levels, including a superintendent, a project executive, a safety person from the office, the design team, and people who rarely go out to sites. They would be trained to go and talk to people and do Gemba walks. 'Listening sessions' were organized, similar to focus groups – a research method that brings together a small group of people to answer questions in a moderated setting. Then, they would come back and talk about what they heard and learned. The ten success factors would be projected on a board and they would investigate which of those factors was covered. Not bad or good, just which one is covered.

The people who were separate from the site but had influence over it, such as the design people and even the project executives who were at the site, were now fully aware of what was happening at the work site. It helped them get a pulse for what was happening more profoundly. This also applied to people from other job sites. They would suddenly realize how other job sites were dealing with similar issues and then would try different solutions to their job site – the cross-pollination of ideas about what is working and not working turned out well. Examples of the success factors are:

- Establish a realistic scope and budget.
- Develop the right margin strategy and align with the contract profit margin.
- Establish a realistic schedule and manage proactively.
- Create a solid operational plan.

DITCHING THE SAFETY BEATING STICK

At one particular job site Ron went to, one of the things he saw right away was the improvement in morale, where it was initially pretty bad. One of the guys from the work floor said, *'I feel like I have been beaten with the safety stick.'* Now – after their improvement efforts as described above – Ron observed the attitudes of people and the kind of interactions between the general contractor and the trade partners and their subcontractors improved. This led to more reporting of issues and dealing with problems. They even saw the incident rate go down eventually.

The organization started to see other improvements, such as a change in the type of leadership, and they noticed a difference in the conversation about what happens when a failure occurs. When Ron started working there, people were scared to report because 'upper leadership will come in, and they will bring the hammer down on us.'

'No one thing is going to change everything; these processes had to happen in conjunction with other elements that were changing because really what we are trying to do is change the mental model of the organization, changing the attitudes and the culture,' Ron explains.

Whereas before, there was an open question of whether the construction company would ever work with the hospital's owner again, Ron's organization had managed to secure and finish the job through their efforts.

WHEN SAFETY AND OPERATIONAL GOALS PARADOXICALLY ALIGN

In the early 2000s, a major airport decided to build a new Air Traffic Control Tower. Only a few issues needed to be accounted for: the new tower would be built in a different place at the airfield, the tower would be twice the size compared to the original, and the new tower would come with new systems and procedures. Essentially, much of the work that the Controllers had been doing daily would change.

What is critical to understand about the work of Air Traffic Controllers is that Controllers are safe and efficient because they have built a particular routine during training and throughout their experience. Routine reduces the workload involved in managing the constant flow of traffic in the air and on the ground and what instructions they want to give the pilots.

During this massive project, a fairly newly hired human factors professional was involved, named Steven Shorrock – Psychologist and Human Factors Specialist. Steven is a well-known and established human factors and safety professional; it is hard to meet anyone who has not met or heard of him. He has many publications to his name, but his contribution to sharing safety information with professionals worldwide is exceptional, as Editor-in-Chief of a Human and Organizational Factors Magazine, as a trainer, and as an associate professor. Back then, since he was new to the project, he was keen to understand the risks of the new operation.

SUBGOAL-CRITERIA WERE MET – BUT THE JOB COULD NOT BE DONE

The training department ensured that the Controllers could work with the new operation. They assessed whether the Controllers could work with the new systems and procedures and whether Controllers had good visibility from the new tower location. The Controllers seemed to pass all these subgoals. Steven found out that a risk assessment had been done where the risks had been described, but he could not grasp how the Controllers experienced the new operation from that assessment. He visited the simulator to see how the Controllers handled the new operation. After four days, he concluded that they could not do their job. The Controllers managed to complete the criteria of the subgoals as defined by the training department, but what they could not do was *handle air traffic safely and efficiently*, their core job. Some Controllers felt so stressed about the new operation that they experienced sleepless nights and worried about the safety of the operation. How did Steven raise these issues with the appropriate organizational levels?

The next day Steven wrote a letter to management. His letter described his observations from the last few days but also included how these observations made him feel. He read his letter to the project leader. The project leader referred Steven to the Head of ATC. Steven went to the Head of ATC and read his letter out loud again. The Head of ATC sent Steven to the general manager, to whom Steven again read his letter, along with several other managers, and had to defend what he observed as the managers were not quite ready to believe him. Besides, wasn't he the new guy?

Steven asked the managers when the last time was that they visited the simulator. It appeared that they had not seen the Controllers work in the simulator in recent months. After this meeting, more meetings followed. Steven explained Hollnagel's (2000) contextual control model during subsequent meetings and asked where everyone thought they stood.

'In reality,' Steven says, 'We were in the midst of chaos and panic, but there was a deadline to be met.' Eventually, the safety director refused to sign off the project's completion until it was assured that the Controllers could do their job. Notice that this is a different type of question than before. Initially, the question was, *'Are the risks mitigated?'*. However, Steven's effort has led to the holistic approach of looking at the operation as more than the sum of parts, and therefore, the question was, *'Can the work be done safely and efficiently?'*.

Having the safety director as the ally did not immediately solve all the issues. There were still concerns about the deadline, which was pushed ahead five months. Project members feared that the airlines and the airport would object to moving the deadline. During a consultation with the stakeholders, however, it became clear that the deadline was not an issue for them. It appeared that both airlines and the airport wanted minimal impact on capacity. Allowing the Controllers more time to get used to the new operation and receive more training would ensure precisely this.

'The bottom line is, get away from behind your desk and start talking to the people who do the work. Then make sure you find a way to have the organization listen to you as well; this is important! I do not want to work for an organization that does not listen,' Steven explains. To achieve this, it would be best to have good persuasion skills, good relationships within the organization, and allies. There is always the chance of optimism bias with the people involved in the project, political games that you might not be aware of, or force fields that might be invisible to you – such as bonuses tied to meeting a deadline. It helps if you are also insensitive to the organization's hierarchy. Even senior management, board members, and CEOs are only human, too, so do not hesitate to knock on their door and make them listen.

EXPLAIN IT TO ME LIKE I'M 5

Listening seems to be a red thread throughout this book. Apparently, it is an activity that brings many benefits, and somehow also an activity that's very difficult to do. To truly understand something, genuine questions need to be asked. Better yet, let's start asking (and I borrow this from a subreddit): *Explain it to me like I'm 5*. Can they explain it? Awesome. Are they struggling to explain it? Perhaps we all have some more exploring to do.

A CLOSER LOOK AT LEADERSHIP BEHAVIOR, TEAM NORMS, AND INDIVIDUAL PERSONALITY TRAITS

'What I have learned – and what the organization has learned – is to *listen*,' says the anonymous Human Performance Expert in the Oil & Gas industry I spoke to.

It is as simple as that. People who feel listened to and cared for perform better. There is much material published about this. Whether it is safety or business performance – they go hand in hand. We all intuitively know that, but the Human Performance Expert says they have studies correlating high employee engagement with improved safety performance. Their leaders invest heavily in 'the soft stuff.' The journey so far has focused on soft skills and seeing improvements as a result. The baseline should always be to listen to the frontline personnel.

What happened within the company for it to reach this realization? The organization surveyed its employees. Throughout the years, they have added questions about speaking up, leaders showing curiosity in work, and questions around trust. Moreover, these are general questions, not questions about safety per se. The result from the survey gives a 'learning mindset score,' a score on psychological safety. These scores reflect whether workers feel listened to.

The organization found that the level of safety performance runs hand in hand with the business performance, which in turn runs hand in hand with the level of employee engagement. This means you can connect the workers' views with the leadership to look for those improvements.

Listening to workers is not always easy for management (and vice versa), especially further up the level. The Oil & Gas industry organization decided to start deliberately and actively listening over time. Listening shows managers care: making time to engage people correctly by asking the right questions.

There are other ways for leaders to get feedback on how they show up as safety leaders. One approach taken by the Oil & Gas organization was for the leader to ask for trusted people who report to them to give structured and honest feedback openly and anonymously. This proved impactful as it sometimes turns out that some leaders are not showing up in the way they believed they were showing up or would like to show up. Then, the journey for the leader starts.

The Human Performance Expert says he remembers a specific conference with a leader on stage. There were about 200 people in the room. This leader was a commercial guy, something like a VP in Marketing. The topic he discussed was safety leadership. He started to tell a story about how outstanding a safety leader he was. The fact that he did such a good job as manager, and things went so well, and he was very proud of the performance. Everyone in the room thought, '*Who is this guy? He is just full of himself.*' And then, at some point, he said he took a moment to think about safety and decided to start asking people. People whom he trusted gave him feedback. At the conference, he started reading out some of the feedback he received. Some of the feedback was quite hard-hitting. He said, '*I am sharing this with you because it made me sit up and pay attention. I was not anywhere near as good as I thought I was.*' That reflection created the starting point for a journey for this leader.

Both efforts allow management to understand how work gets done and realize that mistakes *will* be made. They will know which barriers are weak, appreciate the level of experience within the team, know which processes or procedures are impossible, and know where shortcuts need to be made to complete the work. Realizing this is a starting point for change.

Listening sounds very simple, but to listen with curiosity, to listen empathically, to listen with the aim of improvement, instead of listening to try and prepare your response.

DISARM, LISTEN, CARE

To achieve this level of listening, the Oil & Gas organization spends much time and effort coaching in a non-directive style. They are listening with a minimum number of questions and listening with empathy. People are trained to coach in a non-directive way. Additionally, people are trained in engagement skills. It can be challenging for leaders to start a conversation one step away from the frontline workers.

'So, you have got some guy who is visiting from the UK office, and you would like to chat with a frontline worker who happens to be a Turkish scaffolder who does not speak English; that is very scary for them because they do not know the operation very well, there is a language gap,' the Human Performance Expert says.

These managers must be able to disarm, listen, and care sincerely, and they are trained for these abilities. These soft components are essential, and there are two elements: building a trust-based learning environment at the frontline – a high degree of trust among frontline personnel, and psychological safety is a part of that. The second element is the coaching and engagement from management. Managers have a big drive to master these soft skills to engage with the frontline and get that understanding.

The most senior leaders at the Oil & Gas organization understand the need for this development; however, the middle layers of management face the fundamental dilemmas of production. These managers are faced with the conflicting goals of senior leaders. They are evaluated on their ability to lead as well as their ability to deliver. As in many organizations, sometimes hard delivery dates bite safety goals in the tail. They must manage many things at the same time, from their team to the production, to safety performance, to stakeholders, everybody's time, and down to culture. Here is where the gears can get stuck if one of these elements fails.

PRIORITIZING A LEARNER MINDSET

All leaders at the Oil & Gas organization are expected to adopt a learner mindset and strengthen their ability to create a psychologically safe environment. The noticeable effect of this leadership approach has become visible in the Oil & Gas organization survey: it shows how the leaders are performing in the eyes of the employees. These measures are linked to safety performance.

Work does not begin from the moment someone starts doing the work: work begins much earlier, in the organization, in the system, and in the design. There were some ingrained beliefs about how safety was measured within the organization. The organization made the brave decision to remove the metric Total Recordable Case Frequency from the Group scorecard and also remove it as a target from business and staff goals.

'We found that having this as our primary measure of safety performance was oftentimes driving the wrong behaviors, people trying to fix everything to get all incidents down to zero and losing focus on stopping the most serious life-altering and fatal incidents from happening,' the Human Performance Expert explains.

Getting people to understand why the organization decided to do that has been quite tricky, as it has been believed for several decades that if you can drive down the total number of incidents you are experiencing, the likelihood of a life-altering or fatal incident would be reduced. This belief has been proved wrong. Whereas, if you measure the right things, you can focus on the nasty stuff that has the potential to become a life-changing risk rather than putting your finite effort across everything and anything. They are now measuring the 'capacity of their system' rather than the number of failures. Since then, people have focused their mindsets on the question, *'What does measuring the capacity of our system mean?'* which points more toward meaningful leading indicators such as soft skills.

Down the same line, the Oil & Gas organization has overhauled the way they investigate and learn from incidents. They adopted a consistent causal reasoning model – adapted to focus on human performance analysis. This allowed investigators to ask better questions to get underneath the systemic issues and look for context-driving behavior. The Human Performance Expert says this approach will enable them to 'fail safely' rather than 'fail lucky.'

LIFESAVING RULES MIGHT NOT ALWAYS BE LIFESAVING

Many years ago, the Oil & Gas organization came up with lifesaving rules. The lifesaving rules were distributed in industry, and many organizations adopted them as an industry standard. It is easy for contractors to follow these rules as they are standard throughout the industry. When the organization refined its lifesaving rules, the industry followed suit, and again, contractors faced the same lifesaving rules throughout the industry.

What happens when someone breaks a lifesaving rule? The Oil & Gas organization has seen examples of contractors breaking one of those rules, the person would be marched to the gate and sacked. In the past, the mantra used to be, *'If you choose to break a rule, then choose not to work for this Oil & Gas organization.'* And that was considered harsh. They recognized that that was unfair, as those actions were not malicious. If somebody made a mistake, it is the company's task to uncover in which context that took place and what part of their system had failed.

USING UNKNOWN HAZARDS TO ENSURE DOUBLE-LOOP LEARNING

When the Oil & Gas organization faced a terrible fatality in their operation a few years ago, it concerned a hazard that was not adequately considered. It was not correctly considered because it was deemed so unlikely; the likelihood was off the scale. From a risk assessment point of view, it makes sense to accept this risk. However, it did happen, and the organization experienced a catastrophic event.

The Oil & Gas organization wanted to understand how they might have been caught out. They asked themselves what they could improve within their system to avoid being surprised again by a similar scenario of a hazard that was not considered. They call this their 'outsider-in view of risk,' also called 'double-loop learning.'

DOUBLE-LOOP LEARNING

Double-loop learning is a concept used in organizational development to question and modify existing assumptions and norms, not just solve problems. Most organizations excel at single-loop learning – detecting and correcting errors to achieve established goals. However, they struggle with double-loop learning due to organizational norms discouraging questioning of underlying policies and objectives. To successfully implement double-loop learning, organizations need to focus on valid information, informed choice, and internal commitment (HBR, 1977).

Some parts of the Oil & Gas organization have adopted behavior-based safety (BBS) programs. However, the Human Performance Expert has a problem with this approach as it can end up redirecting the mindset to the frontline workers and their behaviors, seeing them as the problem that needs to be fixed and might miss the real underlying system issues at play. For example, BBS systems may identify that workers are not wearing gloves. The BBS consultant would then ask, *'Why didn't the worker put the gloves on?'* and the answer would be, *'The gloves were unavailable.'* Then, the outcome might be to make gloves available, a box would be ticked, and safety would be resumed. The underlying organizational issue of why the gloves were unavailable in the first place may have been missed.

BEHAVIOR-BASED SAFETY

BBS is an approach to workplace safety that focuses on the behavior of individual employees as the key factor in preventing accidents and injuries. BBS involves 'identifying critical behaviors, observing these behaviors, providing feedback on safe or unsafe performance, and using reinforcement strategies to encourage safe behavior.' This approach is based on the premise that unsafe acts rather than unsafe conditions cause most accidents and that changing employee behavior can significantly impact workplace safety. Other studies have shown that implementing a BBS program can reduce accident rates and improve safety outcomes (Zohar & Luria, 2005).

However, the Human Performance Expert says the premise of BBS is to improve observation skills and create better quality frontline safety conversations, and that is, often missed in the drive to use checklists and gather 'data.' Similarly, some researchers have raised concerns about the potential for BBS programs to shift responsibility for safety away from employers and onto individual workers (Cooper, 2003). While BBS can be a practical approach to improving workplace safety, it should be implemented carefully and with consideration for broader organizational factors (Spigener & McSween, 2022).

OVERCOMING RESISTANCE TO CHANGE IN THE WORKPLACE

One cannot demand people to simply care more. Sometimes, the perception of safety is that it is taking time from doing the actual job. If people need to be trained on all these soft skills like listening and coaching – when are they supposed to get the job done? Nevertheless, the Human Performance Expert says you do not need time as it becomes practice. After coaching training, some questions people ask can make a huge impact. The upfront investment in skill-building can transform someone's ability to have a good conversation. Granted, it requires more than sending someone on a one-off one-hour course or computer-based training module but getting them in a room and practicing for two solid days on applying skills with real people.

CONCLUSION

This chapter underscores the essential role of active listening, open communication, and trust in fostering a culture of innovation and safety. Leaders and organizations can glean valuable insights from the case studies from this chapter to create environments where people feel comfortable taking risks and learning from their mistakes. Consider non-directive coaching and Gemba walks to understand team constraints and challenges. Ask open-ended questions to yield deeper insights and foster a culture of curiosity and learning. Further takeaways are:

- Involve people from all levels of the organization in developing success factors to create a shared understanding of work.
- Foster a learning mindset and psychological safety in your organization. This can be achieved by training leaders in 'soft skills' and showing genuine care for the team members.
- Revisit and revise existing norms and assumptions regularly through double-loop learning. This will help your organization to adapt and grow in an ever-changing environment.

REFERENCES

Cooper, M. D. (2003). Behavior-based safety still a viable strategy. Safety & Health 2003, April, pp. 46–48.

Harvard Business Review (HBR). (1977). Double Loop Learning in Organizations. Retrieved from: https://hbr.org/1977/09/double-loop-learning-in-organizations

Hollnagel, E. (2000). Modelling the orderliness of human action. In Sarter, N. B. & Amalberti, R. (Eds.), *Cognitive engineering in the aviation domain*. (p.65–98). Hillsdale, NJ: Lawrence Erlbaum Associates.

Spigener, L. G., & McSween, T. (2022). Behavior-based safety 2022: Today's evidence. *Journal of Organizational Behavior Management*, 42(4), 336–359. doi: 10.1080/01608061.2022.2048943.

Zohar, D., & Luria, G. (2005). A multilevel model of safety climate: Cross-level relationships between organization and group-level climates. *Journal of Applied Psychology*, 90(4), 616–628.

7 Blending Theoretic Principles to Practical Advantages

All models are wrong – but some are useful.

– George Box, statistician

MODELS IN SAFETY AND RISK MANAGEMENT: AN INTRODUCTION

In safety-centric environments, models can be valuable tools that represent complex systems, identify potential hazards, and implement effective safety measures. There is a wide array of models utilized in safety and risk management, including the bow-tie model, the Swiss cheese model, Heinrich's pyramid, fault tree analysis (FTA), failure mode and effects analysis (FMEA), event tree analysis (ETA), and root-cause analysis (RCA).

The **bow-tie model** visually explains how an incident can occur by identifying its causes, the resulting negative impact, and safety measures (barriers) to manage or avoid potential risks or consequences.

The **Swiss cheese model**, developed by James Reason, depicts multiple layers of defense with holes representing weaknesses. It emphasizes that accidents typically result from multiple failures rather than a single mistake and is predominantly used in the healthcare and aviation industries. When the 'holes of the slices of cheese' line up, it becomes possible that the adverse event takes place.

Heinrich's pyramid suggests that for every significant injury, there are 29 minor injuries and 300 no-injury accidents, highlighting the importance of addressing minor incidents to prevent serious ones.

Fault tree analysis (FTA) and **failure mode** and **effects analysis (FMEA)** are methods for analyzing system and process failures, root-cause analysis, and risk analysis. These are commonly used in manufacturing, nuclear power, and aerospace industries.

Event tree analysis (ETA) is used to evaluate the outcomes of an initiating event, often utilized alongside FTA, while **root-cause analysis (RCA)** is employed across various industries to identify the underlying cause of a problem or accident to prevent its recurrence.

DOI: 10.1201/9781003383109-7

These models offer valuable insights, contributing significantly to improving safety practices. However, they have limitations. All models simplify reality and may not account for all variables or unforeseen circumstances. Their effectiveness also depends on accurate data and careful interpretation.

PATIENT SAFETY: A GROWING PRIORITY IN HEALTH CARE ORGANIZATIONS

We have already met Miriam Kroeze – a Senior Advisor Medical Risk Management at MediRisk – in Chapter 4. Miriam has a compelling story to share which involves the development of patient safety from its onset.

MediRisk was founded as the solution for insuring hospitals for liability, as commercial insurers couldn't insure hospitals, and their insurance was expensive. The hospitals came together and established their own insurance. In the 1990s, nobody had heard of the concept of patient safety. Health professionals accepted the risks of their profession, and it wasn't common practice to discuss incidents or human error.

The Institute of Medicine (IOM) released a report in 2000 titled 'To Err is Human: Building a Safer Health System.' The report stated that errors cause between 44,000 and 98,000 deaths yearly in American hospitals and over one million injuries. Healthcare appeared to be far behind other high-risk industries in ensuring basic safety. The proposition was to break the cycle of inaction after medical errors by encouraging a comprehensive approach to improve patient safety. After this report, the realization came in the Netherlands to also investigate patient safety.

> **Patient safety** is a healthcare discipline that emerged with the evolving complexity of healthcare systems and the resulting rise of patient harm in healthcare facilities. Its focus is to prevent and reduce risks, errors, and harm to patients while providing healthcare. A cornerstone of the discipline is continuous improvement based on learning from errors and adverse events (WHO, 2019).

The Dutch Minister of Health decided that research was needed to gain insight into the status of national patient safety. From this research (Willems, 2004), three significant conclusions were drawn: (1) the Dutch hospitals are not managed based on risks and should start developing safety management systems; (2) the board of directors was not taking responsibility for safety, which was met with surprise by the executives as they deemed the medical specialists to be the experts and that they weren't able to do anything; and (3) the government should also take upon responsibility to enable patient safety. The report led to new research into how often people died or were left with permanent damage from errors in healthcare. This research (de Bruijne et al., 2007) showed that annually 1735 patients died unnecessarily during care in Dutch hospitals. As a result, a five-year program was initiated to work toward improvement in patient safety, divided into ten themes.

Miriam was involved in the program and led the theme of safety management systems. They first organized extensive three-day strategy sessions with the hospitals that included various levels of management, including executives, medical special-ists, Human Resources (HR), etc. They organized these sessions with eight hospitals simultaneously. In that way, all the involved professionals could exchange informa-tion between themselves and between hospitals. They used various methods dur-ing these sessions: dialogue sessions between the hospital executives, the inspecting authorities were invited to join in the discussion, leadership knowledge was used, and Miriam was the chair of these sessions in which she asked questions such as 'How is the patient going to benefit from this proposal?'.

The goal of the three-day strategy session with these hospitals was to produce a plan of approach that all Dutch hospitals could use in improving patient safety. Miriam and her colleagues organized various training and workshop sessions that focused on methods that were at hand, such as incident reporting, the use of bow-ties, and explaining hearts and minds. These efforts led to the initial goal of reducing unnecessary deaths by 50% in Dutch hospitals after five years. The number of 'avoid-able damage' was also reduced.

After five years, the program ended. Later, new research was conducted to gauge the status of patient safety, and Miriam explains that this research indicated that the reduction had plateaued. What was next for the Dutch hospitals? Some hospitals agreed to work together to improve patient safety, share information, and establish a collective safety net, which was undertaken by MediRisk.

What Miriam observed during her time at MediRisk is that over the course of 30 years, the number of claims did not increase, but the costs of medical liability were going through the roof. To illustrate this, in 2013, there was one claim of €850,000, which was really an outlier. However, now they have 27 claims of over one million euro. They discovered that the total cost of claims was caused by extreme claims of over a million euro, and not because of an increased cost per claim. Why was this happening?

HOW HEALTHCARE CLAIMS HAVE INITIATED A POSITIVE CHANGE

Like many other organizations, MediRisk accumulated databases and dashboards with characteristics of the claims for 30 years. The extreme claims, however, were unique as they only happened once in ten years, very similar to a significant accident in other industries. What could they learn from this limited data about significant claims? They did find out that the extreme claims were linked to childbirth – and maternity care where the child had suffered from brain damage, for example. Those types of claims are expensive because of the longevity of the aftercare for both the child and the parents. But there were also instances in which healthcare professionals were involved in a traumatizing event on the job, which led to, for example, post-trau-matic stress disorder, switching jobs, or sick leave. In a nutshell, extreme cases cost money and emotional damage for everyone involved: patients, families, healthcare professionals, hospitals but also for society. How can these extreme cases be pre-vented? What kind of procedures or barriers can be placed? The people at MediRisk realized they needed to start thinking differently about safety because no progress was being made. It would not work to tell people to be better aware during the ten years run-up of a possible extreme claim.

Around the time, the whitepaper on Safety-II was published (Hollnagel et al., 2015), which describes the theory that to enhance patient safety, researchers need to consider complexity in healthcare settings. The whitepaper describes the difference between the two approaches to improving safety.

The first approach – most commonly used in all industries today – focuses on identifying causes and contributing factors to incidents. Incident investigation is a typical example of this approach. The second approach considers variations in every-day human performance and acknowledges that this is why things usually go well. At MediRisk, they asked themselves what they could do daily to increase their reflective capabilities and learn from the things that go well.

Miriam continues, 'In practice, naming what goes well leads to misunderstand-ings. People then think it's about success. Regarding Safety-II, we are interested in the daily routine in which things usually go well and sometimes do not. When you analyze an incident, you see that people have not followed the protocol properly. When you analyze the same case with a good outcome, you see the same thing. The deviation from the protocol in itself is not interesting, but it is why people make adjustments. There's usually a good reason for that. By better understanding the con-text and thus the local rationality, you can also implement improvements that make it easier for people to do the work so that fewer adjustments are required.'

In the healthcare sector, it was standard for all involved healthcare profession-als to work in silos and primarily within their expertise. Failing team performance was one of the most often reported characteristics of the claims of medical errors. MediRisk started to encourage teams to debrief after every birth in a Safety-II man-ner, which meant discussing openly how the multidisciplinary teams thought it went and included the nurses, the gynecologists, the midwives, the pediatricians, and also the parents. They were asked *not* to go through checklists or use any forms but to speak openly about the birth event and finalize the briefing with the question of whether everybody felt good about the delivery. They have now made the debrief-ing more specific to increase the team's flexibility, their resilience. The following questions are essential: were there any surprises during the delivery, and how did we respond to this as a team? The last question remains: *Does everyone leave with a good feeling?* This last question promotes psychological safety in a team and between caregivers and parents.

This turned out to be the most important question, as it created space for care-givers to be honest. Then came out the actual concerns as the trainee would speak up and admit which parts they didn't understand during the delivery, or the parents would ask why everybody suddenly left halfway through the delivery.

By posing the questions in a more Safety-II manner and discussing surprises in which care providers had to adapt, they get a better picture of which adaptations are applied in practice. By building up experience with this as a team, they were able to increase the team's flexibility and resilience.

A Plateau in Growth: Supplementing Traditional Methods for Understanding What's Happening

Jeanette Hounsgaard worked in the chemical industry in Denmark in the 1990s, and at that time, 'zero accidents' were common targets to strive for. At that time, her employer – and many other companies – believed in Heinrich's Iceberg-theory, where at-risk behaviors and latent conditions, near misses, and incidents are at the lower level of the iceberg (below the surface), and accidents are the top of the iceberg and included a ratio of 300–29-1. In Jeanette's company, root-cause analysis was used often, and she realized that somehow they became stuck. In using this method, production was stopped until the root-cause analysis was finished, and the root-cause analysis needed to be done within 24 hours. The reason behind shutting down production was that they didn't know what went wrong, and the same could happen again until they found out and mitigated the risk. The approach was what they could do to prevent the event from happening again.

What was frustrating was that Jeanette saw the same events happening repeatedly, even after implementing corrective measures. Then, she came across Hollnagel's work, which said that you must look at the things that go well, looking into everyday day-to-day work. The idea is that daily work that usually goes well sometimes can go wrong due to variability, and the origin of the events is the same. This approach explains that when an accident occurs, it is not analyzed by something that failed but by the result of the variability in day-to-day work that grew bigger.

> People are more open, honest and willing to participate when you talk about their normal day-to-today coping with varying work conditions.
>
> *– Jeanette Hounsgaard*

Jeanette realized that this might be the missing link. When you talk to the people on the frontline doing the work, they like to talk about how they cope with the challenges in the daily work and different conditions, and they hate to talk about what went wrong. People are more open, honest, and willing to participate when you talk about their day-to-day coping with varying work conditions.

How can you capture the standard day-to-day coping mechanisms? Jeanette had started to use functional resonance analysis model (FRAM) in the healthcare setting in Denmark early 2012. In this job, Jeanette worked at the Danish Regional Centre of Quality – a research and development unit – and was responsible for supporting the hospitals and other healthcare providers to improve patient satisfaction, the quality of care, patient safety, and the effectiveness/efficiency of the services. She gives me an example of the application of FRAM.

When Everything Is Urgent, Nothing Is Urgent

In Denmark, a spine center treats patients who suffer from back pain and need surgery, medical treatment, or physiotherapy. General practitioners can refer patients to specialists. Six doctors do the intake at the spine center, and they can assign urgency to the patient and have a standard procedure to evaluate the urgency, 'urgent' or 'normal.' When she and her team conducted the FRAM, they asked the doctors how they assigned the urgency. After the patient has been assessed and urgency has been given – ranging from normal to urgent, or referring the patient to a different domain – the patient file lands on the desk of the doctor's secretaries. Jeanette and her team then talked to the secretaries.

The doctor's secretaries not only have to face the categories 'normal' and 'urgent' but also 'urgent+' from one doctor and 'urgent++' from another. The secretaries had found their way of interpreting the urgencies the secretary's Doctor A uses 'urgent+,' and that's very urgent, that's more urgent than urgent. Doctor B uses 'urgent++,' the secretaries had to decide which urgency was highest and therefore were responsible for prioritizing the patients. When the doctors discovered this, they realized this approach was not how it should be done, as the doctors were responsible for the prioritization.

What's more, the doctors assumed that the other doctors were following the standard procedures, but they all had developed a slight adjustment to the usual way of working to help the patients. Eventually, they agreed that all doctors would take back the categorization of urgency and all would use either 'normal,' 'urgent,' or 'urgent+' and nothing else. Once a week, they took the same patient, went through the documentation, and decided on a category together. This led to calibration and standardization of assigning patients' urgency and their position in the queue for treatment. This method turned out to be robust in detecting ways of working and making small changes for improvement. What led to the initiation – and acceptance – of using a relatively new tool such as FRAM?

Jeanette tells me that the motivation to use a new approach came after a severe and tragic incident at the psychiatry department in a particular region, where two nurses and one doctor were nearly killed by a patient with a knife. Because of the seriousness of this incident, various authorities got involved in the investigation, including the police, the public prosecutor, and the working environment inspection. The authorities wanted to investigate the cause of the incident and analyzed it with a root-cause analysis. After the investigation, the director of the psychiatry department called Jeanette and her team, saying that the investigation recommended implementing a new procedure.

However, the director felt he couldn't introduce this solution to his staff because it was too simple an answer to a complex situation. The new procedure entailed that the psychiatric patient would have to take off all their clothes to check if they were carrying a weapon such as a knife. The new procedure was dehumanizing for a patient that was in their care. Implementing the new procedure would worsen the relationship between healthcare professionals and the patient. He asked Jeanette if she had a different method of analyzing the incident. After Jeanette and her team used FRAM to investigate what happened, they presented an action plan of 30 areas of improvement to prevent a similar situation.

It took the healthcare organization three years to implement the areas of improvement. Still, it signaled the start of using FRAM in healthcare because the rumor spread that this method brought different perspectives – especially when faced with a complex problem. In this process, people are considered a resource and not a component that could fail by looking at the everyday workflows and explanations and not for causes. Jeanette and her team didn't need to convince other healthcare providers to start applying this approach as the benefits were visible.

DEVELOPING STAFF 'SOFT SKILLS' CAPABILITIES TO IMPROVE PATIENT SAFETY

Initially, personnel in the healthcare sector were mainly recruited based on their medical competencies. More and more attention is being paid to competencies such as teamwork and communication. Discussing these competencies with the people on the work floor was difficult because it concerns attitude and behavior rather than specific aspects of professional competency. They saw that this did not work well at the Leiden University Medical Center (LUMC). There was a turnaround in which the attitude and behavior of staff started to depend on specific themes, and one of those themes is patient safety.

Dr. Jaap Hamming – Chair of Surgery at LUMC – was asked to lead this theme and was not initially enthusiastic about it. This had everything to do with patient safety having a negative connotation because it usually involves complications and incidents or the 'unsafe side' of safety. The LUMC already had experience discussing complications and their surgery's adverse outcomes.

Gradually, Dr. Hamming came across the philosophy of Just Culture and Safety-II, which offered a different perspective. He then discussed with the medical personnel trainers whether this new safety thinking fits within the training programs. There he found that it does work because the medical environment is complex due to the continuous variability of the patient and the situation. Medical professionals' adaptive capacity – such as reacting and anticipating – is enormous and even more important than sticking to protocols. Of course, they need protocols for the practices to be consistent, but this is just the beginning. This means there must be a different way of discussing complications within the organization.

Complications were discussed with the entire surgical department – with all surgeons, such as vascular surgeons, bowel surgeons, and accident surgeons. However, they found that during these discussions, not all the information about such a complication was relevant for all surgeons.

Then, they thought it would be better to have these discussions with the operational teams, so they started the so-called 'quality discussions.' During the quality discussions, everyone within the operations team discussed all cases – the cases that did not go well but also those that did go well. They looked at the previous week's recordings, how they treated them, and whether everyone was happy with them; they checked all the administrative components and whether the registrations of the complications were correct. They also discussed the case when the outcomes have benefited the patient. In addition, they also anticipate the following week, with all patients scheduled for surgery. Were there any hiccups there, were the preparations sufficient, should there be a consultation with a cardiologist? In the past, it has been shown that a patient dropped out of the surgery schedule at the last minute because something was wrong with the preparation and planning.

This approach has been used for five years now, and initially, it took much of their time, about one and a half hours a week. Now, they have reduced that to half an hour a week because the preparations for such a quality meeting have improved, and the patients who will be discharged are prepared by the assistants on the ward. The nurses are also connected to the discussion, which means that all outstanding issues can be discussed immediately and that an action can be developed directly.

EXAMINING THE ROLE OF LEADERSHIP IN PATIENT SAFETY INITIATIVES

The main effect of this approach is that the operations team now feels that they have a grip on their quality, even though they don't have it all on paper. It is due to the consistent and regular reflection with the team on 'the operation,' which makes them feel they have control over it. This approach arose after a conference where Erik Hollnagel talked about Safety-II. As the founder of the concept, they had the theory but not the practical examples. In the evening, they summarized on a beer mat how the idea could be implemented and how the LUMC would apply it. That's how they came up with the system, but it has gradually been adjusted a little each time. They don't discuss what went well; they discuss everyday reality. And what is particularly interesting for them is how the team deals with this, the psychological safety. It is imperative that all operations staff feel safe to discuss everything because sometimes painful things are discussed when someone has made a mistake or misjudged. The Just Culture makes it possible to treat this with respect. The operational employees feel valued; they are asked about their expertise during such a meeting. However, the core is that this is an enlightened discussion and that everyone must have the discipline to do this and thus leave all other activities to be able to attend. Still, everyone has a good feeling about this.

It was more challenging to pitch this approach to management and quality departments. Management would like to see numbers and spreadsheets. Are the complications less, or are there fewer incidents?

'Well, I don't know,' Dr. Hamming says.

'Maybe we have more because we pay more attention to them,' he continues.

The operations team don't have a choice but to keep track of 'the numbers' because management and boards are distanced from the operational departments. If no improvement is visible, then they will question the approach.

'Still, you can ask yourself whether those numbers represent reality. You can discuss what that means. The biggest problem in every sector is a disconnection between management and the operational floor. It also appears more and more that we need to report numbers and more need to talk to each other. If someone from the management or the board wants to know how things are going in the department, they should invite me and ask, "Jaap, how do you do that? How do you view quality and safety in the department?"'

'Then I explain what I'm doing, and immediately tell whether I'm satisfied with it or whether I need something.'

'By the way,' he adds, 'I'm still struggling with that because what does it mean if the trend of wound infections increases after surgery? Do we have more complex patients? Have we had longer surgeries? Are we taking better care of them? The number by itself means nothing.'

That particular discussion has not yet been settled.

FUTURE STRATEGIES FOR ENSURING PATIENT SAFETY AND QUALITY

According to Dr. Hamming, what is the next step in the improvement process? He sees that other departments still have some difficulty with this new way of working, and it is not easy to get people connected. This is because they, as medical professionals, are judged on their production, meaning: operations, patients attending outpatient clinics, and visiting the nursing ward; that's the core work. And everything else – discussing quality – is not work. And that is, of course, quite a strange way of looking at professionalism.

'We used to have these discussions at 5 pm, but it had no chance of success. Of course, everyone wants to go home. You have to do it right when people are fit.

Organizing the care process during the day and the week requires commitment, and we have much to do with that,' he says.

In addition, psychological safety is an aspect that should underlie these discussions. Not all teams and departments have the atmosphere to say what they think. That needs to be actively created. The teams that have good psychological safety also function better. They encourage managers not to manage from their desks. So, they take safety culture measurements, but Jaap says he doesn't look at those. Walking by and talking to each other has more effect.

'Another factor is, how will we make this grip on quality visible in our organization?' Jaap asks. He says he needs a modern safety community, which is not in the classical school of compliance but more on the philosophical side of safety that can indicate what one encounters. It requires creativity but parallels the people with their feet in the clay. Many safety books are written by theorists, and that's all helpful, but those aren't the people on the operational floor.

In addition, they have started to analyze their incidents according to the Safety-II method, and what they have found is that it takes twice as much time to do but provides them with much more information than, for example, a root-cause analysis. The gain lies in satisfaction with those involved – both the patients and the healthcare professionals. That is so intensive that you cannot do it for every incident. They always wonder. 'Is that right?' if they want to do something new.

NORMALIZING VULNERABILITY

Admitting to errors and mistakes is frightening but also liberating. When organizations – such as hospitals – have the courage to admit medical mistakes and what they've learned from them and are open about how they tend to improve, it can help victims in their healing or grieving process. Being this vulnerable requires courage.

'Talking' also, says Bart Jan Jansen (Air Traffic Controller and Supervisor), is the best intervention for safety. He believes in discussing incidents and experiences with colleagues and superiors to normalize the conversation about potential issues.

'Especially about the events in which I was involved. I start with something as simple as, "*Gosh, what happened to me yesterday…*".'

'I do that to make talking about these things a bit normal,' he explains.

As a result, other operational colleagues have also started to share their stories more. The profession of Air Traffic Control is always under a magnifying glass, not only internally but also in society. In the past, one could be seen as an incompetent Air Traffic Controller if he or she was involved in an event, so they never talked about it. Today, it is discussed much more openly. By sharing his experiences, Bart Jan has encouraged others to do the same and learn from each other. He initiated this change in culture to promote learning and improve safety. Additionally, it is vital to establish new procedures and working methods through consultation with operational staff and analyzing data rather than making changes based on isolated incidents.

In any case, what does not work is if a new procedure or working method is introduced without it being explained and established in consultation with the operational staff.

CONCLUSION

Models play a valuable role in safety and risk management across various industries. They are appreciated tools for getting a better understanding of complex systems, predicting potential hazards, and implementing effective safety measures. While all models offer immense benefits, it is crucial to recognize their limitations. All models simplify reality and may not account for all variables or unforeseen circumstances. Their effectiveness also heavily relies on accurate data and careful interpretation. Hence, organizations must exercise thoughtful consideration and understanding while using these models. Organizations striving for improvement could also consider the following:

1. **Continuous learning:** Always strive to keep up-to-date with the latest developments in safety and risk management models. This will ensure that you're leveraging the most effective tools and methodologies in your organization.
2. **Training:** Regularly train staff on how to use these models effectively and safely. This includes understanding the limitations of each model and how to interpret their results correctly.
3. **Holistic view:** Don't rely solely on one model. Use a combination of models to get a comprehensive view of the safety and risk landscape. Different models can provide insights into different aspects of safety and risk, offering a more complete picture.

REFERENCES

de Bruijne, M. C., Zegers, M., Hoonhout, L. H. F., & Wagner, C. (2007). *Onbedoelde schade in Nederlandse ziekenhuizen, dossieronderzoek van ziekenhuisopnames in 2004.* Amsterdam: NIVEL.

Hollnagel, E., Wears, R., & Braithwaite, J. (2015). From safety-I to safety-II: A white paper. doi: 10.13140/RG.2.1.4051.5282.

Institute of Medicine (IOM). (2000). Committee on quality of health care in America. In L. T. Kohn, J. M. Corrigan, & M. S. Donaldson (Eds.), *To err is human: Building a safer health system.* Washington, DC: National Academies Press (US).

Willems, R. (2004). Hier werk je veilig, of je werkt hier niet. Sneller beter – De veiligheid in de zorg. Eindrapport Shell Nederland. Den Haag.

World Health Organization (WHO). (2019). Patient Safety. Retrieved from: https://www.who.int/news-room/fact-sheets/detail/patient-safety

8 Sell Safety Like Hotcakes

Creativity is your intelligence having fun.

– Source unclear

HUMAN-CENTERED THINKING AS THE BASIS FOR SAFETY IMPROVEMENT

Timothy Van Goethem is a Human Performance Expert at SASM and has experience in the pharmaceutical industry. When he started working in industry, he found an organization that kept dealing with the same problem: humans. They kept making the same mistakes. He suggested that it might not be the humans who were the problem but the systems. And that perhaps humans were part of the solution. Most clients are compliance-focused organizations, and Timothy was among the first to ask questions about the compliance-based approach. Pharmaceutical organizations, for instance, must be focused on compliance, on the one hand, because their products are meant to be lifesaving for patients. Timothy says he isn't opposed to compliance and procedures, but he doesn't work well under what he calls 'bureaucratic fascism.'

> The organization kept dealing with the same problem: Humans.
>
> *– Timothy Van Goethem*

In the pharmaceutical industry, the types of near-misses are similar to other sectors: slips, trips and falls, all things 'human error,' and deviation of procedures. These are discovered because operators perform many 'process confirmations,' comparing work-as-imagined and work-as-done. In other cases, workers report near-misses themselves. Thirdly, workers might get caught. The last category is the least informative for the organization. Voluntary reporting poses a problem as well: people who make mistakes get a mark behind their name, and these marks affect their bonuses and chances of promotion. These efforts facilitate workers hiding errors and incidents, depriving the organization of learning.

HUMAN ORGANIZATIONAL PERFORMANCE (HOP): A NEW WAY OF LEARNING

Timothy was motivated to change this way of thinking. He started by determining where the organization wanted to be. The organization's management wasn't blind to the fact that the world is changing and that blame leads to nowhere, but they struggled with how to evolve. HOP emerged, and the company decided to try this out.

Timothy's organization started to look at how they conduct incident investigations and make the process more human-centered. They narrowed it down to '*the human has done something*' – whatever consequence it may have had – and it was somehow

DOI: 10.1201/9781003383109-8

HUMAN ORGANIZATIONAL PERFORMANCE (HOP)

HOP is described as a movement or philosophy using social sciences for a better understanding of how to design resilient systems. There are five principles of HOP:

1. People make mistakes: It's a normal part of being human. Designing systems that can withstand errors prevents injuries.
2. Blame fixes nothing: Thought leaders have long known about the corrosive nature of blame, yet it is still a common reaction to workplace incidents.
3. Context drives behavior: Context is the circumstances that form the setting for an event, such as fatigue, production demands, or broken equipment. For every workplace injury, various circumstances lead to the behavior that resulted in the injury.
4. Learning and improving are vital: The whole point of analyzing workplace injuries is to prevent them from happening again, but many organizations see the same types of injuries over and over.
5. How leaders respond to failure matters: What happens at your organization when things go wrong? Is failure seen as an opportunity or a precursor to an organizational shakedown?

(hophub.org, 2023)

logical to the worker. The question was, *'Can we do something about this uncovered local rationality?'* More information was available to the incident investigators once they started asking these questions, which increased organizational learning.

During one of the operational processes, warehouse items are weighed before being stored to streamline operations. It allows operators to use them without any preparation when needed. However, some elements need to be measured at the time of use, such as the pH value, and if these need readjusting, the operator can add the so-called 'caps': cap 1, cap 2, cap 3, etc.

When the wrong caps are selected – thus bringing in more readjustment than needed – the reactor risks explosion, as they are not built to withstand this new chemical mixture. In one incident, precisely this readjustment occurred, but the operator saw in time that the combination was wrong and was able to halt the process. When the organization looked into why this had happened, they discovered that it was unclear to the operator when the stock items had been weighed and fit to be used in the process and when the items hadn't been measured yet.

It takes a generation to make a lasting change, which is especially true for safety improvements

– Timothy Van Goethem

TICKING THE 'HUMAN-ERROR' BOX

A new human factors training was set up for the incident investigators and operator managers. The focus of this training is on how the process can be improved.

Investigators no longer categorize human behavior in boxes because they've understood it's not about human error, lack of situational awareness, or lack of supervision. The new approach forces investigators to look at the system as a whole. It wasn't 'human error' that made the operators disregard the label of a box – it was the varying interpretability of the label that caused workers to misread it. Rather than recommending *'to be more aware of the label,'* the company changed the labels.

Unfortunately, some habits are difficult to shed. It takes a generation to make a lasting change, which is especially true for safety improvements. In the same company that has been doing so well in adapting to the new way of thinking, some workers still get a mark behind their name once they're involved in an incident. However, it does take much longer before that happens compared to before.

The organization is now at a stage in which they are incorporating human performance in various departments within the organization: safety, quality, operations, and human resources are now integrated. Safety shouldn't be viewed as something stand-alone; it's hardly ever the primary goal of any activity. Organizations shouldn't ignore the complexity of the reality of which safety is only partly the aim.

Does the Worker Fit the Job, or Does the Job Fit the Human?

The organization plans to implement a program that includes safety management and benchmarking. The first step is to declutter by evaluating current procedures, time, and resources to ensure they serve the workers' safety and needs. The focus for the next few years is to communicate the reasoning behind decisions and involve employees in decision-making.

Rather than just telling people they should follow a specific training, involve them and ask whether they think it will be helpful. People are not meant to work as robots and execute orders. Whereas 50 years ago, operators would consist of bakers' sons without any specific training. Today, some operators are highly educated; it would be wrong to hire them and tell them what to do instead of involving them and their creative ideas on how work should be done and with which sources. When the labor market changes, organizations should establish rules and procedures, design the work, and make the job fit to the human.

Selling Safety: Changing the Way We Speak about Safety

Anyone who has seen the movie The Wolf of Wallstreet knows the best way to sell a simple pen. One of the most memorable and insightful scenes shows the protagonist testing his new sales team: sell me this pen. Everyone starts talking about how great the pen is, 'one of a kind'; it writes incredibly well, fits easily in your pocket, and so on. In other words, the focus is mainly on the characteristic properties of the pen. Thus, it's all about knowing how to sell the pen. Only those who understand the art of a sale understand that what you need to be selling is a solution. In the movie, this means needing a pen to write a number down. Similarly, Steve Jobs sold '10.000 songs in your pocket' rather than listing the technical specifications of the new iPod, which not many people tend to understand anyway.

The trick is to package the solution to a problem people might not even know they have.

When it comes to speaking about safety, people tend to lose interest quickly for various reasons, primarily resources. To some, safety is a boring topic that undermines innovation. Or it gets in the way of executing their core business processes that generate profit.

However, people are more likely to pay attention if safety can be effectively marketed as an opportunity for business process continuity. In the end, when safety is somehow compromised, it affects the continuation of the business in various aspects.

It's about preventing business interruption after safety events. Organizations should be able to sell safety to increase customer satisfaction, reduce corporate risk, boost employee morale, and enhance the overall quality of products and services. Management can be persuaded that investing in safety is beneficial for employees and has a positive effect on the business as a whole.

Safety should be considered an integral part of any organization's culture and strategy because it is an investment rather than a cost. Safety could be a way to build brand loyalty and customer trust. Creating an environment where employees take personal responsibility for their safety will help ensure everyone follows safe practices – resulting in higher productivity and less downtime after mishaps.

Selling the idea of safety is a business continuity **opportunity**.

THE POWER OF OPERATIONAL LEARNING: INSIGHTS DIRECTLY FROM THE FRONTLINES

Andrea Baker is a mentor in Human Organizational Performance (HOP). HOP emerged from studying human performance but initially didn't consider the role of leaders and organizations. Andrea aims to support and empower innovators and early adopters in her work rather than forcing changes in people's thinking. She believes change occurs when individuals close to you alter their beliefs or behavior.

During a consultation process, she and her colleagues look for individuals in the organization to help them reach out to others. For instance, after speaking with a plant manager, she and the manager engaged with the employees on safety concepts and theories, their assumptions about management, human error, and its preventability. They give people time to ponder these ideas before presenting them with stories and studies that clarify a direction. Once they have established a theoretical understanding, they move on to operational learning, teaching people how to put these ideas into practice. Her approach involves providing some structure, but primarily she teaches them how to have conversations differently than before.

'In complex systems, it's crucial to prioritize psychological safety,' Andrea explains. Multiple everyday occurrences going wrong simultaneously can lead to failures. To improve operational learning, it's essential to comprehend the daily routine and tasks from the worker's viewpoint. Excitingly, facilitating operational learning at the organizational level enables individuals to share their learning experiences from various sites or levels, leading to centralized resources to tackle organizational issues. Consequently, organizations evaluate their management practices to ensure they align with the learning process.

Instead of breaking everything down, making small adjustments to certain elements is better. For instance, if audits aren't providing helpful information for improving organizational performance, the questions asked can be modified to encourage more engagement and learning. Other tweaks might include changes to HR policies, responses to certain events, or how procedures are written. However, these changes may take 4–5 years to implement fully. HOP encompasses many domains, such as safety, change management, leadership, and business management. HOP enables organizations to shift their focus where it's needed. Andrea and her partners work with organizations to assess their progress and determine areas requiring more attention. One aspect where HOP has shown its value is in data input.

If an organization is resistant to change, shifting the focus to individual members may be helpful. Andrea suggests there is no monolithic 'organization' but rather a group of individuals. The critical question is which individuals are ready to embrace change. This could be a small or large number, and it may only take one person. Andrea's role is to identify individuals open to change and provide them with the necessary tools to make the shift. It's not about forcing change upon those who are resistant. Once early adopters have experienced the benefits of change, they can share their stories with others and encourage them to follow suit.

> Changing paradigms requires respect and politeness, as it opposes what people have known and used for years. A natural reaction to that is frustration.
>
> – *Andrea Baker*

The efforts made through HOP result in noticeable improvements based on the information gathered from past events. Comparing an event before and after changes in thought processes, the actions taken to improve the organization are significantly different. Often, before adopting the HOP mindset, organizations provide counseling, reiterate existing procedures, or implement secondary checks into existing procedures after learning from operational experiences. After the shift, they lean toward addressing weaknesses in process design, addressing brittleness in areas where it is easy to fail and focusing on improving mitigation controls.

Understanding the operational context and creating focus areas for the organization has proven successful in various aspects: from the severity of injuries to the financial consequences of worker injuries. More intangible impacts are the way that people work with each other, how easy it is to get information throughout the organization, the type of interaction people have by just changing the way some of the types of questions are asked, and the realization that it's not the leader's job to dictate solutions for issues, but to (1) learn from the workers in order to help define the issues, (2) enable the workers to try out solutions within the workers' sphere of control, and (3) own the elements of the problem that fall within the leaders' purview. Also, today's solution might work today but not tomorrow. That means that the interaction needs to be high at all times, and people should remain open to ideas for change and solving issues together.

WHY IS IT SO DIFFICULT TO CREATE MORE 'AWARENESS'?

I shared an experience with Andrea about one of the companies I worked for; the executives asked the safety department to organize a company-wide safety awareness training – from HR to the operational personnel and all departments in between. I specifically remember asking my manager to ask the executive team to clarify what they wanted precisely. Creating 'more awareness' seemed like the executives didn't really know what they wanted, but something had to be improved. Andrea says these types of requests are very recognizable and tells me what she would do through the HOP lens: define what the intent is of the training.

> Inattentional blindness is a human condition that can't be fixed; therefore, 'increasing awareness' seems impossible.
>
> – Andrea Baker

Andrea had often found that the intent is based on some misunderstanding about how people operate and the belief that if more 'awareness' through training is created, people will pay more attention. The key is to explore whether this is true by talking to the executives and finding out what they are trying to accomplish. If they have seen an increase in slips, trips, and falls and want people to pay more attention to this, Andrea would discuss the context of the work in which these events occur.

Slips, trips, and falls will always be around. The organization should look for where these events would have dire consequences for workers and then try to mitigate the circumstances to minimize the risk of dire consequences. Ask the employees who do the work to teach you in what places they are working that would create these dire consequences after a slip, trip, or fall. Here, you learn what could be done differently; it's specific to certain areas and not a 'blanket approach' to the whole organization.

It is often no different when proposing new ideas, concepts, and tools.

SUCCESS IN NAVIGATING CHANGE IS ONLY AS GOOD AS YOU SELL IT

Many people start with a new organization in good spirits. They are hired based on their competencies, experiences, and attitudes. Yet that energy quickly dissipates when they find that the intended changes are not shared with the rest of the team, the ideas are nipped in the bud, or their efforts are going nowhere. All of this can lead to frustrated employees.

However, you can also take a closer look at things. Nicklas Dahlstrom, whom we met in Chapter 5, says to be realistic because he also sees this often happen in people.

'You have to look at what you can change, no matter how small. Also, look at the type of organization and realize to what extent they will accept change. Finally, you would do well to see it as a long-term project.'

'Your idea may be as good, but if you don't know how to sell it, you won't get it. It's a shame, but it's not how the world works. People don't change because someone tells them to change or to do something different. What you want is to make people look at things differently,' he says.

THE POWER OF TELLING STORIES IN SAFETY COMMUNICATION

Tony McConachie – Safety Innovation Transformation Lead – worked as a Manager Safety Strategy & Systems at a rail company in Australia. He has a background in organizational development and change management, mainly in HR. He used to be critical and judgmental about safety and didn't want to work in that field because he thought it was challenging and irrelevant to operational people. However, he was now in charge of safety strategy in his company.

He had previously held a position in change management at a public water management organization. During this time, he worked with a safety culture partner to initiate small changes focusing on operations. Together, they explored and introduced new safety theories, such as Safety-II and Safety Differently, which he found intriguing. Fortunately, his organization was dedicated to developing and improving safety thinking, which involved bringing in experts and conducting various masterclasses and sessions with external thought leaders. This created a sense of momentum for him, and his role was to develop a new health and safety strategy, along with various initiatives that aligned with the new view of safety. He also supported the team and operations in implementing and embedding this new strategy.

'The first thing we had to do, was to understand the problem. The previous safety team was very, very separate from operations. There was no connectedness, and there was no value adding,' he explains.

'We needed to understand what worked well and what kind of sucked about the approach to safety. We used a human-centered design approach to make changes, to undertake a whole series of interviews and focus sessions with key executives, the CEO, health and safety representatives, and everyone involved in health and safety in that particular process,' he continues.

'First of all, we needed to gather the people's voices and understand what we need to keep doing that they like and what we need to do to cut out the things that don't help them in their jobs.'

'Like many other organizations, we had that "zero harm" slogan. We ripped all of those up. We all know that's a bit of a joke; it drives the wrong type of culture. We needed to get that out of the way. That was part of our strategy that our executives signed off. We didn't want a slogan, and we didn't want to replace it with anything else.'

The methods used to investigate incidents were not yielding positive results. Root-cause analysis, TapRoot® investigations, and similar methods were heavily focused on blame and administrative aspects, resulting in many disciplinary letters attached to safety investigations. The investigations lacked the ability to enhance safety measures and often resulted in numerous safety indiscretions. It was suggested to detach HR from safety investigations to improve the process. Despite opposition, efforts were made to improve the investigative process and achieve better outcomes.

To implement their strategy and initiatives, the team needed approval from the executives and CEO. They faced some challenges with the legal counsel and some managers who believed disciplinary action was the only way to hold people accountable. However, this approach went against accountability and instead promoted blame and punishment.

The team created a learning methodology for investigating incidents and gaining knowledge about day-to-day work, including near-misses. This approach was effective because it provided a consistent experience for everyone, regardless of whether something went wrong. Initially, some team members were hesitant, but after a year, they came to appreciate its value. They also developed report templates and a dashboard to support the Learning Teams' approach, which improved communication and strengthened operational learning. The methodology significantly impacted operational staff, who were highly receptive to it.

Operational staff appreciated the shift from the outdated safety observation method where individuals were labeled as safe or unsafe. This process was judgmental and unpleasant, with leaders observing and criticizing workers. Instead, they created the 'work insights process,' which focuses on critical control work. It can be customized for both critical control and normal work activities.

In practice, their initiative facilitated employees to engage with their peers to discuss high-risk activities. Each month, they would focus on a specific activity and send employees to where the work is happening to capture insights on how the job is conducted. These insights are captured by peers, leaders, and the safety team and added to their system.

Previously, the organization relied solely on numerical data. However, they've since implemented a new approach where they capture specific sections of the report that highlight key stories. These sections consist of brief paragraphs that cover the top six or seven stories based on the month's high-risk focus. In addition to numerical data, they now incorporate storytelling to make the report more engaging. The graphs and numbers are still included, but the focus is on presenting the information in a more interesting way.

SIMPLIFYING SAFETY DOCUMENTATION FOR SMS IMPROVEMENT

Next, Tony's organization planned to improve its safety management system (SMS). They wanted to tackle the Holy Grail process of pre-task risk assessments.

'We had this awful three-page checklist fest approach that everyone hated. Either no one really did it, or they filled it out in bulk and just pulled them out of the truck,' Tony explains.

A group of field workers were gathered in a room and asked to review a checklist process displayed on the wall. They were given red pens and were told, *'I want you to cross out the things that really annoy you and just leave the stuff that you would still want in there.'*

They had a list of about 30 questions and checkboxes, but only a few were important to their job planning and risk assessment. The group narrowed it down to seven guided questions and worked together to customize the process. They shifted from a heavily documented approach to one with zero documentation and a simple seven-question process. This process was then added to every work order and job issued to the frontline workers. They could still refer to the questions if needed, but they could easily navigate the process and answer the questions once familiar with it.

The operational personnel highly valued the change program and considered it to be an exceptional experience.

GETTING RID OF THE 'GOLDEN RULES'

As part of their strategy, the organization eliminated the 'golden rules' and didn't replace them with anything else. Tony says they were used to be copied and pasted into disciplinary letters and saying, '*Gotcha, you broke that rule, and you broke that part of the Code of Conduct.*' He edited the safety section of the Code of Conduct and removed the blamey language when it came up for review as well.

The health and safety representatives could spend four times more time with the team. This resulted in significant operational improvements observed by the Operational Executives and General Management. The team's job turnaround time improved, and the need to repeat processes decreased. Eventually, the team was able to focus on the important safety measures that truly protect people instead of getting distracted by unnecessary safety noise that hindered their work.

A NEW PERSPECTIVE ON AUDITING

Tony and his colleagues also examined their auditing approach thoroughly.

'Traditionally speaking, auditing practices are absolutely embarrassing and woefully ineffective. They are very judgmental and closed in their process. I really dislike most auditing,' he says.

To comply with legislative requirements, it is important to understand how the work is being carried out within the system. Gathering stories from those experiencing the process is crucial, as it may not always go as expected due to its non-linear nature.

'You actually need to spend a lot more time understanding and being empathetic and curious and engaging and your style with people,' Tony explains.

Tony reported a decrease in serious near-misses and injuries after implementing incremental changes. The previous approach incentivized employees to report as many hazards and near-misses as possible, with financial rewards for those who reported the most each month. However, this strategy resulted in low-risk incidents being overlooked. The new approach focuses on high-potential events and has increased safety surveys and psychological safety. Additionally, the company has seen an increase in morale, engagement, and enablement during a time of restructuring.

GENERATIONAL CHANGE: THE LONG ROAD TO LASTING SAFETY IMPROVEMENTS

Meet Ben Goodheart, a safety expert who began his career working in aviation in 1993 when safety measures were not as advanced as they may be today. He later ventured into safety in other high-consequence industries ranging from coal mines and paper mills to healthcare and public utilities. He asked numerous questions to better understand work in those organizations, especially concerning daily operations that carry a high penalty when they fail. The pivotal moment for him was often when he asked, 'How do you do this so well every day and make it work?' He believes that people make sense of the world through stories, and that asking genuine questions about work allows rich stories to emerge.

> When you retire, how will someone else know how to do your job as well as you do? To be as good in your job as you are?
>
> – Ben Goodheart

In one particular organization, Ben was on a safety project where a worker was responsible for checking a valve containing liquid chlorine. He followed all the safety training he was given by wearing protective gear such as eyeglasses, gloves, hearing protection, and proper clothing.

After isolating the valve, he removed his eye protection and gloves per his training. However, despite the valve indicating no pressure, there was a failure in the system that trapped pressure inside, causing liquid chlorine to be released and endangering the worker. He collapsed near the valve but was luckily rescued by a coworker and recovered afterward.

He was fired.

He was fired for not following a procedure from 1985 that had never been updated and was unknown to almost everyone in the facility. However, he had followed his training and wore eye protection, which made it almost impossible to see the valve well. All workers were trained to remove their eye protection and gloves to see better and perform the job effectively. This created a dilemma as it was impossible to both follow the procedure and do the job well. If you look hard enough, you'll find someone to blame. The organization failed to inquire about the workers' job requirements and the challenges they encountered while carrying out their tasks.

> It's not about understanding policy; it's about the human connection within a system. The problem is that if you look hard enough, you can always find someone to blame after an adverse event.
>
> – Ben Goodheart

Instead of the story having an unhappy ending, the organization uses it as an example of its maturing view of safety. The worker's termination was reversed, and a formalized learning process was undertaken, during which the stories from various employees helped make sense of what happened. Those stories provided context for action and for lasting safety evolution. That process was counter to over 100 years of working, but it is a story that is now embedded in the culture of the organization.

At the request of executives, Ben and the team at Magpie Human Systems have conducted ethnographic safety climate studies in various companies. These studies – reliant in part on structured interviews and sensemaking tools that help uncover and understand behaviors – usually result in reducing ineffective or inappropriate procedures instead of adding new ones. Ben mentions that stories and storytelling are only really useful to the extent that we're also engaged in story-*learning*. That process is where ethnography and sensemaking are crucial.

'We're handing over the ability – with some structure – for folks to help others learn from their experiences, and that's a massively powerful safety tool,' Ben says. As a result, the mindset of chalking up accidents to the need for operational personnel to have 'more training' or 'pay more attention' is eliminated or at least reduced in favor of a fuller understanding of the context of work.

Ben believes translating these stories to management does not have to be challenging. He has spent considerable time communicating with senior management and executives, sharing stories about how operational personnel make sense of their work. Goodheart's experience and research show the importance of learning from stories, but humans also have transformative learning experiences when they create their own stories as a participant. Ben and his team have taken these managers on-site to observe and understand what their teams do to create successful daily work. They have spent shifts – just like workers do – without cellphones, brought their own lunch, been trained on tools and equipment, and worked alongside the operational personnel in aircraft, trains, line training facilities, electrical utilities, and maintenance trucks. They work overnight shifts, work out in the weather, and use stories as a sensemaking tool to learn how people adapt the way they work so that changes in safety are done *with* people doing the work and not simply *to* them. Managers may have hesitated to do this in the past, but increasingly, they are pleased to see the work being done and take pride in it – especially when they understand the results.

CONCLUSION

Prioritizing safety in organizations might be challenging, as management faces many stakes, processes, and stakeholders. In this chapter, we've taken the cases of safety professionals to establish some strategies that have worked for their organizations. We've looked at how organizations view human error, the way jobs are designed, and applying various models to manage safety, such as the Viable System model. Furthermore, we've proposed that selling safety is much like marketing for other industries: It's how you frame the proposition of safety being a vital element of business continuity. To be able to do this, understanding the core business process and the possible impact it might endure during disruptions is crucial. Finally, we embrace all views on safety thinking and discussed a rather controversial paper that challenges the New Safety view. In the final chapter of this book, we will dive deeper into actionable ways to assess risk, safety, safety culture, and possible impact on business continuity.

REFERENCE

Hophub.org. visited on 31st August 2023.

9 Chance Favors the Prepared

Change before you have to.

– Jack Welch

SAFETY MAY BE SUBJECTIVE; RISKS ARE REAL

Safety may be experienced as subjective, varying from person to person based on their perceptions, experiences, and circumstances. What one person considers safe, another may view as unsafe. However, despite the subjectivity of safety, risks remain an objective reality.

As I conclude this book and reflect on the shared stories, I realize that organizations must recognize that safety thinking is continuously developing. By aligning ever-changing safety practices with business continuity, organizations can stay ahead in mitigating risks; preventing accidents; and safeguarding their employees, assets, and reputation. Additionally, despite continuous debate about the topic of safety culture, industries in general accept that a robust safety culture forms the foundation for safety and, thus, for business continuity.

Wrapping up all the experiences I've heard when interviewing professionals for this book, I conclude that, before initiating any safety program, organizations would benefit from:

1. Determining which metrics are sensible to keep a measure of – discard or automate the rest.
2. Using both incidents as well as normal work to learn.
3. Involving relevant personnel by understanding in the conversation.
4. Having the courage to be vulnerable.
5. Approaching safety and business process continuity holistically.
6. Understanding how their core business may be impacted if safety is compromised.
7. Ensuring they've identified a comprehensive list of risks and mitigations.

WHY WE SHOULDN'T ONLY BE TALKING ABOUT SAFETY BUT ALSO ABOUT BUSINESS CONTINUITY

Organizations can become more resilient if they prioritize safety as part of their business continuity plan. Business continuity plans are in place to avoid disruption of operations, and being adaptable, mitigating risks, and quickly recovering from disruptions go hand in hand with safety management. There are organizations, and parts of organizations, that view safety as an assumed precondition of the operation

DOI: 10.1201/9781003383109-9

that requires effort only after an adverse event. Talking about business continuity when talking about safety efforts transcends compliance-driven efforts in planning for safety. Considering business continuity when talking about safety plans requires various levels of management to have a seat at the table – not just the safety manager.

CHANGE YOUR SAFETY GAME: MEASURE THE ORGANIZATION'S PLANNING FOR SAFETY CULTURE

One tool that can be used to gain insight into what the organization is already doing well and where the potential for improvement lies is the AVAC-SCP, which was introduced in Chapter 3. This tool was developed by the Amsterdam University of Applied Sciences – Aviation Academy (Piric, et al., 2018) to offer practical guidance and a comprehensive framework for developing and enhancing organizational safety culture.

Recently, the tool has undergone further development to align with the needs of the industry. The improved tool – the Safety Culture Conditions Tool – provides organizations with practical guidance to develop and nurture a robust safety culture. With its list of 37 carefully curated prerequisites, organizations can ensure that they cover all essential aspects of safety culture development. These conditions encompass the entire spectrum, from documentation to implementation and employee perception.

One of the standout features of the tool is its scoring method, which enables organizations to monitor performance and make meaningful comparisons between different departments. By systematically evaluating and scoring each element of the safety culture prerequisites, organizations can identify strengths, weaknesses, and areas for improvement. This data-driven approach empowers organizations with actionable insights.

To get a feel for what the tool investigates, see Table 9.1. Note that this is a short version of the actual tool and only to be used to familiarize with the method rather than as an analysis method. For more information on the use of the tool, visit www.safety-rebels.com.

TABLE 9.1

A short excerpt from the Safety Culture Conditions Tool (www.safety-rebels. com)

Policy	Implementation – for managers	Perception – by employees
There is a written commitment of management toward safety.	My commitment toward safety is clearly visible.	The organization encourages me to be committed toward safety.
The organization facilitates a questioning attitude (e.g. peer reviews, brainstorm sessions, formalized feedback).	I encourage a questioning attitude (e.g. peer reviews, brainstorm sessions, formalized feedback).	In my experience, I can voice my questioning attitude within the organization.

(Continued)

**TABLE 9.1
(Continued)**

Policy	Implementation – for managers	Perception – by employees
There are documented definitions of 'unacceptable' and/or 'acceptable' behaviors.	I use documented definitions of 'unacceptable' and/or 'acceptable' behaviors to evaluate someone's behavior.	The organization is clear about what 'acceptable' and 'unacceptable' (safety) behaviors are.
Rights and duties of employees about safety occurrences are described.	Rights and duties of my team regarding safety and safety occurrences are known to them.	It is clear what rights and duties employees have regarding safety occurrences within the organization.
The organization recognizes that there may be a gap between rules and regulations and operational processes.	I recognize that there may be a gap between rules and regulations and operational processes.	In my experience, there is a gap between the rules and procedures and the way me and my team actually do the work.
Authority is delegated to employees that allows them to self-organize their work within set boundaries.	I delegate authority and self-organization of the work to my team when needed.	When needed, I have the authority to do the work how I see fit, and I am able to self-organize the work.
There is a policy for safety reporting.	My team reports safety occurrences, even the small ones.	I always report safety occurrences, regardless of their severity.
It is documented that all reporting is non-punitive.	As a manager, I ensure reporting is not punished.	In my experience, reporting of safety occurrences is not punished within the organization.
A safety information system is in place.	Me and my team frequently use the safety information system.	Me and my colleagues frequently use the safety information system.
There is a policy for sharing safety information across the organization through safety activities (safety meetings, workshops, etc.).	I share safety information across my team through dedicated safety activities (safety meetings, workshops, etc.)	Safety information is shared with me and my colleagues through dedicated safety activities (safety meetings, workshops, etc.)
The need to learn from safety failures (e.g. safety investigation reports, voluntary reports, audits) is recognized.	I use information from safety failures (e.g. safety investigation reports, safety audits, voluntary reports) to improve learning.	The organization uses information from safety failures (e.g. safety investigation reports, safety audits, voluntary reports) to improve learning.
Company policy urges the organization to also examine successes relative to an incident during safety investigations.	During safety investigations, I also examine successes relative to the incident.	The organization is also interested in the successes relative to an incident during safety investigations.

CONCLUDING REMARKS

In my search for practical approaches to make positive safety changes throughout various industries, I've spoken to over 30 professionals who have inspired me (again) about safety. Some of the methods are well-known, others are relatively new. I have never intended to depict these stories as the ultimate truth – they only serve as examples of what has worked for the professionals who shared them. Every organization is unique, and it is an illusion to think that there is one way, method, or concept that will provide the solution to all its safety challenges, no matter how scientifically proven. The journeys organizations have embarked on are compelling, and the professionals I've spoken to have agreed to share them with you as the reader. This was my intention: To share their stories.

REFERENCE

Piric, S., Roelen, A. L. C., Karanikas, N., Kaspers, S., Van Aalst, R., & De Boer, R. J. (2018). How much do organizations plan for a positive safety culture? Introducing the Aviation Academy Safety Culture Prerequisites (AVAC-SCP) tool. *AUP Advances*, 1(1), 118–129.

Index

Note: **Bold** page numbers refer to tables and *italic* page numbers refer to figures.

AcciMap 53
accountability
    just culture 18
    leadership 63–65
    safety performance indicators 46–47
Air Navigation Service Providers 18
air traffic controllers (ATC) 14, 16, *18,* 22, 67
Air Traffic Control Tower 76
Analysebureau Luchtvaartvoorvallen
        (ABL) 19
Anglo-Saxon approach 28–29, *29*
auditing approach 101
Aviation Academy Safety Culture Prerequisites
        (AVAC-SCP) 38–39, *39,* 105
aviation laws, Article 5.3 20
awareness 97–98

behavior-based safety (BBS) programs 81
bow-tie model 83
business continuity plans 104–105

Civil Air Navigation Services Organization
        (CANSO) 33
communication
    cross-cultural 24
    open 11, 18, 61
    opportunities for improving interpersonal
        73–74
    training program 73
company-wide safety awareness training
        97–98
complex problems, sense-making framework 33
compliance-focused organizations 93
complicated problems, sense-making
        framework 33
construction company 72–74
constructive leadership 8
corrective leadership 8
crew resource management (CRM) 21, 22
crisis management 30
cross-cultural communication 24
cross-industry insights 69
Cynefin model 33

decision-making process 33, 44
    employees role in 95
    local rationality 56
Delta Airlines incident 14–15, *15*

double-loop learning 80–81
Dutch Safety Board (DSB) 15, 16

Edmondson, Amy 25
effective safety leadership 59–60
event tree analysis (ETA) 83

failure mode and effects analysis (FMEA) 83
fault tree analysis (FTA) 83
flexible culture 35
flight data monitoring (FDM) 49–51
focus of safety
    from safety metrics to proactive safety 49–50
    types of *44*
functional resonance analysis model (FRAM) 48,
        49, 64, 87, 88

Gemba walks 73–75, 82

healthcare claims 85–86
health care organizations, patient safety in 84–85
Heinrich's pyramid 83
high-consequence tasks 6
hindsight bias 45
Hofstede's cultural dimensions *23,* 24
human-centered thinking 93–97
Human Organizational Performance (HOP) 94,
        96
human performance principles 10

ICAO *see* International Civil Aviation
        Organization (ICAO)
incident investigation 86
    goals of 54–55
    learning from incidents process 56–57
    local rationality in development of event
        55–56
individualism *vs.* collectivism 24
indulgence *vs.* restraint 24
informed culture 35
Institute of Medicine (IOM) 84
Integrated Energy Company 6, 7, 9
intensive agriculture 30
International Civil Aviation Organization (ICAO)
        33
International Labour Organization (ILO) 34
International Nuclear Safety Advisory Group
        (INSAG) 34

just culture 18, 25, 26, 35, 90
    empowering employees with growth and
        63–65
    philosophy of 89
    principles of 18

key performance indicators (KPIs) 43

lagging indicators 43–44
laissez-faire leadership 8, 60–61
leadership
    effective safety leadership 59–60
    emergent behavior from societal
        developments 63
    listening 77–78
    operational and commercial goals 62
    operational experience 61–62
    patience, persistence, positivity 68–69
    role in patient safety 88–90
    servant 63–65
leading indicators 43
learning
    from incidents process 56–57
    just culture 18
learning culture 35
Learning Team methodology 70
Leiden University Medical Center (LUMC)
        89, 90
lifesaving rules 80
listening 79
    disarm and care 79
    leadership behavior, team norms, and
        individual personality traits 77–78
    prioritizing learner mindset 79–80
litigious society 52
local rationality 55, 56
local vs. global interventions 53
long-term vs. short-term orientation 24
loss aversion 21, 22
lost time injuries (LTI) 51

MediRisk 48, 84–86

Occupational Safety and Health Administration
        (OSHA) 34
Oil & Gas industry organization 78–80
oil slick approach 3–4
open communication 11, 18, 61
operational goals
operational learning, power of 96–97
operational workplace environment 72
organizational learning 5, 6, 45, 55, 94
organizations
organogram 28–29, 29

patient safety 84–85
    developing staff soft skills capabilities 88–89
    role of leadership in 89–90
    strategies for ensuring quality 90–91
performance-shaping factors (PSFs) 9, 10
persistence 68–69
pharmaceutical industry 93
positional leaders 8
positive safety culture 34
positivity 68–69
power distance 24
psychological safety 25, 91

Regulation (EU) No 376/2014 19
reporting culture 35
Resilience Engineering 56
resistance 50–51
Restorative Just Culture approach 45
Rhineland approach 28–29, 29
risk management 20
    models in 83–84
risk-taking behavior 10–11
root-cause analysis (RCA) 83, 87

safety assurance 20, 42
safety-centric environments, models in 83–84
safety communication, power of stories 98–100
safety culture 33–34
    first rule of 36
    of five elements 35
    ladder 35–36
    measurement and limitations 37
    planning prevents poor performance 38–39,
        39
    problems with 35
Safety Culture Conditions Tool **105–106**
safety dashboards 47
Safety-I 55
Safety-II 44, 46, 54, 55, 57, 69, 86, 89, 90, 91
    philosophy of 89
safety improvements
    generational change 101–103
    human-centered thinking as basis for 93–96
safety leadership 8, 10
    principles 7–8
safety management system (SMS) 18, 20, 25,
        33–34, 38, 42, 100
safety performance indicators 50, 52, 55, 58
safety policy 20
safety practices 12
safety promotion 20
safety-specific transformational leadership 61
Schiphol Airport 14–16, 15
servant leadership 8, 63–65

simple incident method (SIM) 56
simple problems, sense-making
        framework 33
situational awareness 9–10
Sjursøya train accident 59
SMS *see* safety management system (SMS)
subconscious behavior 52–53
subjectivity of safety 104
Swiss cheese model 83

Tenerife accident 22
Tipp-ex incidents 47
total recordable injury rate (TRIR) 43
training interventions 61
transactional leadership 8

transformational leadership 8, 60, 61
trust 10–11, 63
    just culture 18
    relationship between safety and 10, 11
trust-building activities 11

UK Civil Aviation Authority 65
uncertainty avoidance 24
unsafe behaviors 65
unstable approach 57

Work-as-Done 38
Work-as-Imagined 38
workers' safety and needs 95
working communities 31

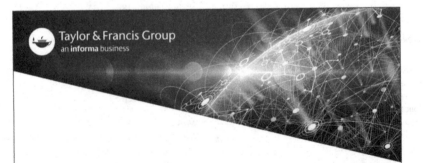

Printed in the United States
by Baker & Taylor Publisher Services